JOURNAL OF CYBER SECURITY AND MOBILITY

Volume 3, No. 4 (October 2014)

JOURNAL OF CYBER SECURITY AND MOBILITY

Aim
Journal of Cyber Security and Mobility provides an in-depth and holistic view of security and solutions from practical to theoretical aspects. It covers topics that are equally valuable for practitioners as well as those new in the field.

Scope
The journal covers security issues in cyber space and solutions thereof. As cyber space has moved towards the wireless/mobile world, issues in wireless/mobile communications will also be published. The publication will take a holistic view. Some example topics are: security in mobile networks, security and mobility optimization, cyber security, cloud security, Internet of Things (IoT) and machine-to-machine technologies.

Published, sold and distributed by:
River Publishers
Niels Jernes Vej 10
9220 Aalborg Ø
Denmark

Tel.: +45369953197
www.riverpublishers.com

Journal of Cyber Security and Mobility is published four times a year.
Publication programme, 2014: Volume 3 (4 issues)

ISSN 2245-1439 (Print Version)
ISSN 2245-4578 (Online Version)
ISBN 978-87-93237-65-0 (this issue)

JOURNAL OF CYBER SECURITY AND MOBILITY COMMUNICATIONS

Volume 3, No. 4 (October 2014)

Foreword

It is great pleasure to introduce you to the October 2014 issue of Journal of Cyber Security and Mobility. This is issue number 3 and Volume 4 of the series of this journal. This issue consists of three papers from open call. The first paper by Lars R. Knudsen considers the concept of Dynamic encryption that is all about cryptosystems for secrecy. In dynamic encryption approach the sending party can choose the encryption system at random from a set of many, secure crypto systems, encrypt the clear text and transmit the result together with some additional information. In this system the receiver does not know anything a priori except for the value of the secret key. Knowledge of the secret key is sufficient for the receiver to decrypt the cryptogram and retrieve the message from the sender. This cryptosystem has many applications including electronic mail systems, mobile conversations, and cloud storage. Second paper by Josef Noll et al. titled, "Measurable Security, Privacy and Dependability in Smart Grids," presents a methodology for assessing security, privacy, and dependability (SPD) of the embedded systems. This methodology uses a multi-metrics approach to evaluate these SPD values during the running processes thereby allowing optimisation towards a balanced solution. The authors have analysed three different use cases including billing, home control and alarm and have validated its applicability by implementing it in a real use case for a smart grid scenario. The third and final paper titled, "Study on Estimating Buffer Overflow Probabilities in High Speed Communication Networks," is authored by Izabella Lokshina. In this paper the author proposes new methods to estimate the probability of buffer overflow in high-speed communication networks. Since the buffer overflow is a rare event, the authors have used rare event simulation approach with Markov chains to conduct the analysis.

We would like to thank the members of the steering board, advisors, reviewers of the articles and colleagues from River Publishers for their efforts towards the production of this issue. We hope the readers will enjoy these articles and we solicit contributions and guest editors for future issues of the journal.

Editors-in-Chief
Ashutosh Dutta, AT&T, USA
Ruby Lee, Princeton University, USA
Neeli R. Prasad, CTIF-USA

Dynamic Encryption

Lars R. Knudsen

DTU Compute Technical University of Denmark
lrkn@dtu.dk

Received 15 January 2015; Accepted 2 February 2015;
Publication 3 April 2015

Abstract

This paper considers the concept of dynamic encryption[1], which is about cryptosystems used for secrecy.

1 Introduction

Traditionally the participants in a secure communication scenario involving encryption need to agree on the particular algorithm used for the encryption and they need to establish a shared, secret key that is known only to the legitimate parties. Kerckhoffs' principle says that a cryptosystem should be secure even when attackers know everything about the system except for the value of the secret key (in a particular session). Here we introduce cryptosystems for which the receiver, Bob, does not know anything *a priori* about the system except for the value of the secret key, but without necessarily violating Kerckhoffs' principle and without sacrificing on security. The scenario is in more details: Alice and Bob exchange a secret key by some protocol. Alice (alone) decides on the cryptosystem to use, encrypts the cleartext with the agreed key and sends the cryptogram (and some additional information) to Bob. Although Bob does not know the cryptosystem used, the knowledge of the secret key is sufficient for him to decrypt the cryptogram and retrieve the message from Alice. The proposed method has many practical advantages and applies well to for example electronic mail systems, mobile conversation and

---

[1] Patent pending

Journal of Cyber Security, Vol. 3, 357–370.
doi: 10.13052/jcsm2245-1439.341

cloud storage. Moreover, the transmitting party in a conversation can change the cryptosystem as often as desired, e.g., for every new communication. This means that an attacker will have a harder time breaking the security, since he would first have to figure out how the encryption was effected, and thereafter he could try to break the encryption. This supports the strategy of "moving-target defense" that is becoming popular.

Encryption systems have existed for many years, several thousands of years if we are willing to believe the historicans. There are two main types of encryption, the symmetric-key encryption and the public-key encryption. The typical scenario of symmetric-key encryption is as follows, see also Figure 1. The sender, usually called Alice, and the receiver, usually called Bob, first agree on a key-exchange protocol and a particular encryption mechanism or encryption algorithm. Then they exchange a secret key K, such that after the execution of the protocol only Alice and Bob know the value of the secret key. Subsequently, Alice and Bob can exchange encrypted messages. In a public-key cryptosystem Alice and Bob each have a pair of keys, a private key and

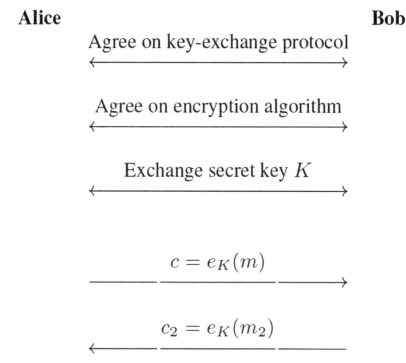

Figure 1   Typical scenario for symmetric-key encryption

a public key. The public key is made available for everyone and the private key is known only by one party. Then Alice uses Bob's public key to encrypt a message, and Bob can decrypt the message using his private key. Bob can send Alice a message by encrypting it with Alice's public key and Alice only can decrypt the message using her private key.

To evaluate the security of an encryption system it is customary to make at least two assumptions.

1. It is assumed that an attacker has access to the cryptograms sent between the sender and the receiver.
2. It is assumed that the attacker knows all details of the encryption process except for the value of the secret key.

The second principle is known as an interpretation of Kerckhoffs' principle [7, p. 225]. Kerckhoffs' principle is that the security of an encryption system should lie in the secrecy of the key and not in the secrecy of the system. Many examples from history (e.g., the World War II) show that this is a sound principle.

Today Kerckhoffs' principle is standard in cryptology-security evaluations. One advantage by using only public systems is that people will evaluate them and try to break them. A system which is kept secret can only be evaluated by the possible few people who knows about it.

2 New Approach

The new approach presented here can very shortly be described as follows: The receiver does not need to know any details of the encryption process except for the value of the secret key. Here the sender will choose the encryption system and encrypt in such a way that the receiver does not need to know how the encryption was done and still be able to decrypt correctly. And there are variants of this approach that do not have the disadvantage which comes from violating Kerckhoffs' principle.

In principle the new approaches can be applied to both symmetric-key and public-key encryption systems, but we shall concentrate on the former.

There are several ways of implementing the above idea, in this text we give two such examples.

2.1 Wrapped Encryption

It is assumed that the sender and receiver have agreed upon a secret key K. The sender does the following:

1. Choose (at random) a symmetric-key encryption system with encryption function E and decryption function D.
2. Construct an algorithm Ω which on input the secret key K and some ciphertext, implements D and returns the plaintext.
3. Compile Ω to the executable code X.
4. Send X to the receiver.
5. Encrypt the plaintext m using E and the key K and save the ciphertext c.
6. Send C to the receiver.

The receiver does the following:

1. Execute the code X.
2. Type in K and c on request.
3. Save plaintext m and delete X.

This algorithm is suitable for many different scenarios, including any data transmission e.g., email, voice transmission and cloud storage. There are applications where it may be advantageous to combine the transmission of X and c into one step. In this case, c can be incorporated into Ω. The compiled code X would then contain the ciphertext, which would not have to be transmitted separately. See Figure 2 for an example in the C programming language.

2.2 Encrypted Algorithm

It is assumed that the sender and receiver have agreed upon two secret keys k_1 and k_2, and that they have agreed upon an encryption system ε. The sender does the following:

1. Choose (at random) a symmetric-key encryption system with encryption function E and decryption function D.
2. Encrypt D using ε with the shared key k_1 and transmit the result to Bob.
3. Encrypt the plaintext m using E and the key k_2 and transmit the ciphertext c.

The receiver does the following:

1. Decrypt the first part of the received ciphertext using k_1 to retrieve D.
2. Run D on the second part of the received ciphertext using k_2 to retrieve m.

In many applications, it would be acceptable that the encryption system ε is not very fast, since it is used only to encrypt a relatively short string. This would allow to choose a very strong cipher for ε, e.g., triple-AES.

```
#include ''stdio.h''

int ciphertext={''aeni92fj!~&k''};

void decrypt(int *ciphertext,int *key, int *plaintext)
{....

 plaintext = ....
}

main()
{
 printf(''Type in your key '');
 scanf(&key);

 decrypt(ciphertext,key,plaintext);

 printf(''Plaintext is .... '',plaintext);

 exit(0);
}
```

Figure 2   C-program example for data transmission

2.3 Discussion

Note that this approach does not conflict with using Kerckhoffs' principle. As an example, assume that there are n secure symmetric-key systems. Then the sender above can choose one of these systems at random for every message she sends. The receiver does not need to know which algorithm has been used to encrypt a particular plaintext. In §3 below we discuss some ways of constructing several symmetric-key systems from existing ones.

The main advantages and disadvantages of the approaches are the following:

Advantages

- *Adds another element of secrecy:*
  Since neither the receiver nor the attacker knows the cryptosystem which has been used for the encipherment, a successful attacker would need to first retrieve the encryption algorithm used, then try to break it. Moreover,

if the transmission between Alice and Bob is executable code, an attacker would have to decode the executable code into a higher level language in addition.

- *Adds another element of security:*
  Attacks on symmetric-key cryptosystem often require a huge amount of inputs and outputs from the encryption algorithm. If the encryption of many and/or long plaintexts is split between several secure encryption systems, one cryptosystem will be used to encrypt fewer plaintext blocks and the chance of a successful cryptanalytic attack will decrease. This supports the strategy of "moving-target defense" that is becoming popular [1, 4].

- *Adds efficiency to the system:*
  The receiver does not need to know the decryption algorithm, so this removes the need for the sender and receiver to first negotiate which encryption algorithm to use.

Disadvantages

For both approaches there will be some additional setup time in the system, and for the wrapped encryption approach:

- Transmission of executable code: users may be reluctant to execute program code received from other party. There are several ways to deal with this.

  - The sender can use a data integrity mechanism and send checkcode along with the other data. The receiver will check whether the received code is authentic before processing it. The integrity check should depend on the secret key. This is known as *symmetric-key authentication* and there are several international standards for this.
  - The code is executed in a protected programming environment. This would limit the risk in case of hostile code.

3 Customization

In this section methods we discuss how to construct private versions of existing encryption systems.

3.1 Variations of Existing Algorithms

Some encryption systems have been designed to allow for customizations, e.g., by choosing some of the components in the design according from a large set of primitives. As an example consider the *wide-trail strategy* behind the design of AES [8]. As for most symmetric encryption systems the AES can be split in some linear mappings and some nonlinear mappings. In the wide-trail strategy these two sets of mappings are constructed independently according to some predefined sets of constraints. For any components satisfying these constraints the result is a secure encryption system, where "secure" here is relative to the predefined constraints.

The estimated security levels of modern encryption systems are often calculated by assuming that the subkeys used in each iteration are independent. However, often the subkeys are computed from the shorter user-selected key in a so-called key-schedule. Further customizations of a system can be obtained by modifying the key-schedule.

3.2 Module Encryption

Also, one can use what we will call "module encryption". Most, if not all, symmetric-key encryption systems used in practice are so-called iterated ciphers. This means that the ciphertext is computed in a number of iterations, also called rounds, as a function of the plaintext and the secret key. In *module encryption* one constructs a number of iterated ciphers, say *s*, each with a relatively small number of rounds. Take the secret key K and generate from it a number of subkeys to be used in the small ciphers. Assume that the concatenation of t such small ciphers each dependent on a different subkey yield a strong cipher, resistant against all known attacks. In this way it is possible to select s^t variant ciphers.

3.3 Cascade Ciphering or Multiple Encryption

In a cascade of ciphers

$$\varepsilon_r \circ ... \circ \varepsilon_2 \circ \varepsilon_1$$

the encryption result of the first cipher ε_1, is encrypted again using the second cipher ε_2 and so. This is also called *multiple encryption* when the ciphers ε_i are similar.

It has been shown that for all attacks exploiting plaintext statistics a cascade of ciphers is provably at least as secure as the first cipher in the cascade [6]. Under a chosen plaintext attack the cascade of ciphers is at least as secure as any cipher in the cascade [2].

As an example, if ε_1 would be AES, any cascade of ciphers would be at least as secure as AES. Therefore one can implement dynamic encryption which is as secure as AES, yet the receiver and the attack do not know how the encryption was computed. Since AES is known, this implementation would not have the security risk one gets (traditionally) by violating Kerckhoffs' principle.

4 Variants of the AES

First we give a specification of the AES with the notation that we will use. The AES is an iterated cipher which runs in 10, 12, respectively 14 *rounds* depending on the size of the key of 128, 192, respectively 256 bits. These variants are named AES-128, AES-192, and AES-256.

The AES uses four main operations in a single round. In the following, let r denote the total number of rounds applied in encrypting a single block, thus for AES-128, $r = 10$. We use G_i, $1 \leq i \leq r$, to denote the round function which takes a 128-bit block as input and provides a 128-bit block as output. The ith round for $I \leq i \leq r-1$ is defined as[2]

$$G_i = \text{AddRoundKey}_i \circ \text{MixColumns} \circ \text{ShiftRows} \circ \text{SubBytes}.$$

The final round of an AES encryption is special since MixColumns is omitted

$$G_r = \text{AddRoundKey}_i \circ \text{ShiftRows} \circ \text{SubBytes}.$$

Before the first round, a pre-whitening key is used in a step AddRoundKey_0, so the r-round encryption with master key k is denoted by

$$\text{AES} - 128_k = G_r \circ \cdots \circ G_1 \circ \text{AddRoundKey}_0.$$

Each of the four operations operate on a 128-bit block arranged in a 4×4 matrix over the finite field $\text{GF}(2^8)$ defined via the irreducible polynomial $x^8 + x^4 + x^3 + x + 1$. In this finite field, an element is represented by a single byte $a = (a_7 a_6 \cdots a_1 a_0)$, where $a_i \in \text{GF}(2)$, which in turn represents the field element

$$a(x) = a_7 x^7 + a_6 x^6 + \cdots + a_1 x + a_0.$$

We use hexadecimal notation to write byte values. For example $a = 01$ represents the polynomial $a(x) = 1$ and $a = 02$ represents $a(x) = x$, and so on.

---

[2]The notation $g \circ f$ is function composition of f and g. The input to the composition is evaluated through f, the result is then fed to g and the output from g is the final result

In the following, we briefly describe the four operations used in an AES round. The text to be encrypted is 128 bits which is arranged as 16 bytes in a 4×4 matrix.

4.1 SubBytes

In the SubBytes operation, each of the 16 bytes in the state matrix are replaced by another value according to an 8-bit lookup table, an Sbox.

$$\begin{pmatrix} a & b & c & d \\ e & f & g & h \\ i & j & k & l \\ m & n & o & p \end{pmatrix} \rightarrow \begin{pmatrix} S(a) & S(b) & S(c) & S(d) \\ S(e) & S(f) & S(g) & S(h) \\ S(i) & S(j) & S(k) & S(l) \\ S(m) & S(n) & S(o) & S(p) \end{pmatrix}.$$

For decryption one uses the inverse Sbox, which is easy to compute.

4.2 ShiftRows

In the ShiftRows step, the ith row of the state, $0 \leq i \leq 3$, is left rotated by i positions: ShiftRows:

$$\begin{pmatrix} a & b & c & d \\ e & f & g & h \\ i & j & k & l \\ m & n & o & p \end{pmatrix} \rightarrow \begin{pmatrix} a & b & c & d \\ f & g & h & e \\ k & l & i & j \\ p & m & n & o \end{pmatrix}.$$

For decryption one uses right rotations instead of left rotations.

4.3 MixColumns

In this step, each of the four columns of the state matrix are multiplied from the right onto an invertible matrix M in the finite field. The matrix M is

$$M = \begin{pmatrix} 02 & 03 & 01 & 01 \\ 01 & 02 & 03 & 01 \\ 01 & 01 & 02 & 03 \\ 03 & 01 & 01 & 02 \end{pmatrix}.$$

For decryption one needs the inverse matrix which is

$$M^{-1} = \begin{pmatrix} 0e & 0b & 0d & 09 \\ 09 & 0e & 0b & 0d \\ 0d & 09 & 0e & 0b \\ 0b & 0d & 09 & 0e \end{pmatrix}.$$

4.4 AddRoundKey$_i$

The $r + 1$ round keys, denoted $rk_0, \ldots, rk_r$ are generated using the AES key schedule, cf. later. In this step, the 128-bit round key, rk_i, is added bitwise modulo two (using XOR) to the state.

The AddRoundKey operation is the same for encryption and decryption.

4.5 The AES Key Schedule

The round keys in AES are viewed as matrices with elements in the finite field GF(2^8). The first pre-whitening key rk_0 is the n-bit master key itself, so $rk_0 = k$. The key schedule varies slightly across the three AES variants. Here, we describe it for AES-128 and refer to the literature for the other two cases.

We consider the four columns of the two round keys as $rk_i = (rk_i^0 || rk_i^1 || rk_i^2 || rk_i^3)$ and $rk_{i+1} = (rk_{i+1}^0 || rk_{i+1}^1 || rk_{i+1}^2 || rk_{i+1}^3)$. To derive rk_{i+1} from $rk_{i,}$, $0 \leq i < r$, we do the following

1. Let $rk_{i+1}^j = rk_i^j$ for $j = 0, 1, 2, 3$,
2. Rotate rk_{i+1}^3 such that the byte in the first row is moved to the bottom,
3. Substitute each byte in rk_{i+1}^3 by using the Sbox from the SubBytes operation,
4. Update the byte in the first row of rk_{i+1}^3 by adding 02^{i-1} from the finite field, and
5. Let $rk_{i+1}^j = rk_{i+1}^j \oplus rk_{i+1}^{j-1 \, mod \, 4}$ $for\ j = 0, 1, 2, 3$.

This procedure is repeated for $i = 1, \ldots, r$ to obtain the round keys $rk_0, \ldots, rk_r$.

4.6 Variants of the AES

There are several possible ways of making variants of the AES without destroying the ideas behind the design. The subfunctions in the encryption process are AddRoundKey, MixColumns, ShiftRows, and SubBytes.

AddRoundKey. The generation of round keys can be modified in many different ways without inviting to new cryptanalytical attack. However, it is not a trivial task to figure out how many secure ways there are of doing this.

MixColumns. This mapping is implemented by a circulant 4×4 matrix over GF(2^8). According to [3] there are around 2^{31} possible ways to choose such a matrix with properties similar to the one chosen in the original AES.

ShiftRows. There are $4! = 24$ possible ways of choosing this mapping.

SubBytes. There are $2^8! \approx 2^{1684}$ possible bijection on eight bits, so there is plenty to choose from. In practice, one would probably prefer to use a (relatively) short key to pick one of these permutations, e.g., with a 128-bit "Sbox-key" we would pick one of 2^{128} permutations. Therefore we would only get a relative small subset of all permutations but we would still want to sample them uniformly.

To choose such an S-box one can use the Knuth shuffle which is simple and intuitive [5]. The algorithm uses random numbers in the range [0, i] for varying values of i, which may not so easy to compute. An alternative is to use the initialization part of the stream cipher RC4, developed by Professor Ron Rivest from MIT [9].

5 A Dynamic Encryption Variant - Using AES

In this section one variation of dynamic encryption using the AES is presented.

We introduce a new cipher, RAES, which is similar to AES-128 except that the S-box used is different. RAES takes two keys each of 128 bits. One key is used just as the key in AES-128 is used, the other key is used to select the 8-bit S-box used in RAES. AES-128 has 10 rounds, cf., earlier, but RAES can be specified with less than 10 rounds.

Now we can define a dynamic encryption variant using AES.

DynAES

$$\text{RAES}_{dk3,dk2}\text{AES}-128_{dk1} \circ \text{AddKey}\,[dk_0].$$

This is an AES encryption using dk_1 with a so-called prewhitening key dk_0, followed by an encryption using RAES, where dk_2 is used as the "AES-key" in RAES, and dk_3 is used to select the 8-bit S-box.

For decryption one goes in the reverse direction using the inverse Sbox in RAES, which is easy to compute.

5.1 GenerateSbox

There are many possible ways of generating an 8-bit S-box. We have chosen to use the key setup function of the stream cipher RC4.

The input to the function is an s-byte key $y = (y_0, \ldots, y_{s-1})$.

- let T be initialised such that $T(i) = i$ for $i = 0, ..., 255$
- set $j := 0$
- for $i := 0$ to 255 do

  $- \mathrm{j} := \mathrm{j} + \mathrm{y}_{i \bmod s} x + K[i \bmod s] \bmod 256.$

− swap($T(i)$, $T(j)$))

The output is the table T.

The table T which is output from the key setup function can be as the Sbox in an AES variant. In the example above we used s $= 16$.

5.2 Security of the Variants

It is possible to prove that the above AES variant is at least as secure as the AES itself, but likely much more secure.

Theorem 1 *The cryptosystem DynAES where the keys dk_0, dk_1, dk_2 and dk_3 are chosen independently at random is* **at least as secure** *as the AES-128.*

Also note that this proof is valid regardless of the number of rounds of the cipher RAES, also if RAES runs in zero rounds.

6 Conclusion

This paper has presented the approach of dynamic encryption. The sending party in a conversation can choose the encryption system at random from a set of many, secure cryptosystems, encrypt the cleartext and transmit the result together with some additional information. The receiver will be able to decrypt the received cryptogram on input the correct key. The receiver does not need to know how the encryption was performed, what is important for the receiver is to do the decryption and retrieve the message. The dynamic encryption approach has many applications.

References

[1] ACM/SIGSAC. First acm workshop on moving target defense (mtd 2014). http:// csis.gmu.edu/MTD2014, November 2014.

[2] Shimon Even and Oded Goldreich. On the power of cascade ciphers. In David Chaum, editor, CRYPTO, pages 43–50. Plenum Press, New York, 1983.

[3] Otokar Grosek and Pavol Zajac. Searching for a different aes-class mixcolumns operation. In *Proceedings of the 6th WSEAS International Conference on Applied Computer Science, Tenerife, Canary Islands, Spain,* pages 307–310, 2006.

[4] Homeland Security. Moving traget defense. http://www.dhs.gov/science-and-technology/csd-mtd. Retrieved November 26, 2014.

[5] Donald E. Knuth. *The Art of Computer Programming, Volume II: Seminumerical Algorithms, 2nd Edition.* Addison-Wesley, 1981.

[6] Ueli M. Maurer and James L. Massey. Cascade ciphers: The importance of being first. *J. Cryptology*, 6(1):55–61, 1993.

[7] A. J. Menezes, P. C. Van Oorschot, and S. A. Vanstone. *Handbook of Applied Cryptography.* CRC Press, 1997.

[8] National Institute of Standards and Technology. Advanced encryption standard. Federal Information Processing Standard (FIPS), Publication 197, U.S. Department of Commerce, Washington D. C., November 2001.

[9] R. L. Rivest, 1996. Attributed to Rivest in *Applied Cryptography* by B. Schneier, Wiley, 1996.

Measurable Security, Privacy and Dependability in Smart Grids

Josef Noll[1,2], Iñaki Garitano[2], Seraj Fayyad[1,2],
Erik Åsberg[3] and Habtamu Abie[4]

[1]*University of Oslo, Oslo, Norway,*
[2]*UNIK, Kjeller, Norway,*
[3]*eSmartSystems, Halden, Norway,*
[4]*Norwegian Computing Centre (NR), Oslo, Norway*
{josef, igaritano, seraj}@unik.no; erik.aasberg@esmartsystems.com;
habtamu.abie@nr.no

Received 1 December 2014; Accepted 2 February 2015;
Publication 3 April 2015

Abstract

This paper presents a methodology for assessing security, privacy and dependability (SPD) of embedded systems. The methodology, developed through the European collaboration SHIELD, is applied for the smart grid network as deployed in the South of Norway. Three Smart Grid use cases are analysed in detail, being billing, home control and alarm.

The SHIELD methodology uses a Multi-Metrics approach to evaluate the system SPD level during running processes and compares it with use case goals for S, P, and D. The simplicity, applicability, and scalability of the suggested Multi-Metrics approach is demonstrated in this paper. It shows that a single configuration is not sufficient to satisfy the given goals for all use cases.

Keywords: Smart Grid, Security, Privacy, Dependability, Embedded Systems, Internet of Things, Measurable Security, Advanced Metering Infrastructure, AMI.

Journal of Cyber Security, Vol. 3, 371–398.
doi: 10.13052/jcsm2245-1439.342

1 Introduction

Our society is built and driven by Embedded Systems (ESs), ranging from low-end systems, such as smart cards, to high-end systems, like routers and smart phones. ESs thus constitute one of the key elements of the Internet of Things [1]. The technological progress produced several effects, such as the power and performance boost of ESs. Hence, their capabilities and services have raised, and in consequence, their usage has been substantially increased.

Together with the evolution of performance, energy consumption and size, ESs jump from isolated environments to interconnected domains. Although the evolution of connectivity enlarges the number of possible services, at the same time it increases the *attackability* of this kind of systems. When isolated, ESs were hard to attack, since attackers need to have physical access. However, the open connection towards Internet makes them vulnerable to remote attacks.

ESs are used for multiple purposes, mainly to capture, store and control data of sensitive nature, e. g. home or cottage usage. Attackers could have different goals to compromise ESs, from gathering sensitive data, thus compromising their privacy, to disrupt the service by a Denial of Service (DoS) attack, exploiting their security and dependability. The consequences of a malicious and a successful attack could cause physical and economic losses, and thus it is important to keep them as secure, privacy-aware and dependable as needed in a given situation.

In this paper, a functional SPD level evaluation methodology is presented and its applicability validated by implementing it in a real use case for a smart grid scenario.

Traditional smart grid installations focus on measuring the power consumption of the home. As such a smart grid infrastructure provides opportunities for the power-grid provider and incentives for the end-customer in saving power under high-demand circumstances. Extending the smart-grid infrastructures with home-control and alarm functionalities can open novel areas of operation for smart-grid providers. Our paper presents a smart-grid operation in the South of Norway, and the security demands of the operator towards novel services.

The core of the methodology resides in the Multi-Metrics SPD evaluation, which provides a practical and simple solution for SPD implementation during not only the design, but the whole lifetime of ESs. The presented concepts and results are developed through the European activity SHIELD. The nSHIELD project [25] looks at the applicability of the envisaged approach in different domains. This paper focusses on selected use cases for the smart grid, including billing, alarm and home control.

The rest of the paper is structured as follows: Section 2 shows a current and future view of Smart Grid; Section 3 provides an overview of related work on security, privacy and dependability; Section 4 introduces the smart grid scenario and explains the selected use cases; Section 5 presents the ESs SPD level methodology; Section 6 describes the evaluated use cases and describes the metrics used in this paper; Section 7 introduces the Multi-Metrics approach and shows its applicability by analysing the SPD level, SPD, of a the smart grid sub-system; Section 8 evaluates the results obtained in the previous section and finally, Section 9 provides a summary of the key contributions of this applied research work.

2 Smart Grid System and Services

Since Whyte published his patent on the powerline communication system back in 1975 [33], a variety of methods have been implemented to monitor the power consumption in a building. These methods range from low end optical readers of the numbers of the analogue reader to high-end control systems for the home, and are commonly treated as automatic meter readers (AMR). Karnouskos et al. described the change from an AMR system (AMS) into an advanced metering infrastructure (AMI), including initial security challenges [12].

This work introduces measurable security for AMI, focussing on the extension from meter readings into the home control. A typical infrastructure is indicated in Figure 1, collecting the meter readings in a concentrator and then using the mobile network to connect to the control centre.

Though a variety of security papers have been published regarding AMI, e.g. the work from Beigi et al. on intrusion detection [3] and the work from Saputro and Akkaya on privacy in smart grids [27], little is published combining various aspects of security. Our approach is based on the industrial applicability of measurable security in an existing smart grid infrastructure. Our main focus is to see to what extend the deployed infrastructure (see Figure 1) can satisfy the needs of advanced services, including:

- Monitoring the grid to achieve a grid stability of at least 99, 96%,
- Alarm functionality, addressing both the failure of components in the grid, as well as alarms related to the Smart Home, e.g. burglary, fire, or water leakage,

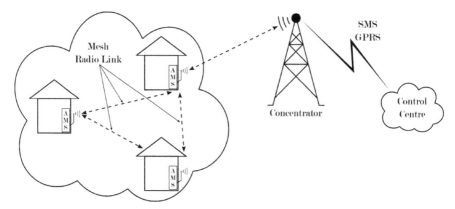

Figure 1   Smart Grid for automatic meter readers

- Billing functionality, providing at least the total consumption every hour, or even providing information such as max usage,
- Remote home control, interacting with e.g. the heating system,
- Intrusion detection, monitoring both hacking attempts to the home as well as the control center and any entity in between, and
- Fault tolerance and failure recovery, providing a quick recovery from a failure.

Examples of SPD analysis in the smart grid include for **S**ecurity the control unit of a home, or the hacking of the control center; for **P**rivacy the habit monitoring of the residents, e.g. nobody is at home; and for **D**ependability the supply security of the electrical power grid, which is mainly the component dependability of the grid. Further details on these selected use case are provided in Section 4.

3 Related Work

Security, Privacy and Dependability SPD and specially their measurement have been analysed through several papers. This section describes a representative set of those papers, most of them analysing SPD aspects individually, without considering all together at all.

Security, Privacy and Dependability metrics can be classified into *(i)* system-based and *(ii)* attacker-based measurements. On one hand, system-based metrics, also called as system-centric approaches, concentrate on system components and capabilities [22]. On the other hand, attacker-based measurements, or attacker-centric approaches, assume attacker capabilities, resources and behaviour. Previous research on security measurement has been mainly concentrated on attacker-centric approaches [30, 31], even if there are some system-centric methods [22]. When considering Smart Grids, the system-centric approach considers the design and configuration aspects rather than attacker capabilities and behaviour.

Several papers have been published considering security, privacy and dependability for Smart Grid. However, few of them measure the SPD level of a system by combining different metrics.

Referring to security, Mo et al. [24] analyse different Smart Grid security requirements by highlighting attack models and possible countermeasures. Due to the lack of security metrics, the work highlights the importance of identifying a set of metrics that combines and addresses the security concerns during the design of secure Smart Grid. In a similar way, Wang et al. [17, 32] explore Smart Grid security objectives by analysing possible attacks. Furthermore, they analyse network and cryptographic countermeasures and they suggest the design of secure network protocols.

Focusing on privacy, it is important to highlight that the exploitation of consumption data obtained from individual houses could end in a severe privacy violation. The knowledge obtained from the analysis of consumption data has a high economical value where different sectors could be interested [5]. The information gleaned through metered energy data processing can be demonstrated with the use of non-intrusive appliance load monitors (NALM), which can recognise and track appliance usage patterns [15, 26]. In this sense, Cavoukian et al. [4] analyse several privacy aspects to consider during the design phase. They classify the features into end user equipment, electricity distribution and generation. Kalogridis et al. [11] suggest an approach to mask the electricity usage by providing to the Automatic Measuring System (AMS) a balanced electricity consumption. To this end this work suggests the installation of a rechargeable battery and some power mixing algorithms. Furthermore their work proposes three different metrics to evaluate the privacy level of the end user real and *faked* measurements, *(i)* the relative entropy, *(ii)* the similarity based on cluster classification, and *(iii)* a regression analysis.

Regarding dependability metrics in Smart Grid, there are few papers in the literature using different metrics to evaluate dependability. One of them is the work of Gungor et al. [7], which uses LQI and RSSI metrics to evaluate the quality of a radio link. RSSI is the estimate of the signal power while LQI is used as chip error rate. The main purpose of this work lies in the characterisation of radio links for the usage of Wireless Sensor Networks (WSN) in Smart Grid. In the same way, Li et al. [16] use delay, cost and path length metrics to measure the Quality of Service (QoS) of Smart Grid packets routing. To continue with dependability metrics, the work of Gungor et al. [8] analyses different aspects of Smart Grid communication and highlights the necessity to measure the reliability of the communication system. This work is not going further with metric definition, and instead of measuring the reliability as a combination of multiple metrics, it defines reliability as a metric. The work presented by Lauby [14] is focused on the measurement of the energy distribution reliability. In this sense they define two different metrics, the System Average Interruption Frequency Index (SAIFI) and the Customer Average Interruption Duration Index (CAIDI). The first metric measures the number of sustained outage events experienced by a end customer, while the CAIDI metric refers to the average length in time of a end customer outage.

There are few research activities which propose a kind of SPD metric in the field of Smart Grid. Furthermore, the previously analysed ones are focused at most on three different metrics. Their objective is to measure the security, privacy or dependability levels of specific system components or to define a broad metric which involves several components but, without going far into the details. The methodology presented in this work defines and evaluates several metrics together to end up with a common SPD level result.

In addition to Smart Grid, the measurement of security, privacy and dependability have been analysed in different fields such as software. Considering the publications which combine multiple metrics resulting in a common measurement, Howard was one of the first to introduce an attack surface metric [9], which has been an starting point of multiple publications [2, 10, 13,19, 23, 28–29] for measuring the software security in different domains. Defined as the *attack opportunity* or *attackability* of a system, or its exposure to attack [9, 10], attack surface is a relative metric that strikes at the design level of a system. One of the essential ideas behind attack surface metric is that it is important to remove unnecessary features, and offers those characteristics as reconfigurability or *composability* options.

Howard et al. [10] propose the attack surface metric for determining whether one version of a system is more secure than another with respect to a fixed set of dimensions. Their work evaluates the attack surface metric of five different versions of Windows operating system. To do so, they define and use five different elements to evaluate the Attack Surface level; *Target, Enabler, Channel, Protocol and Access rights*. After giving a specific weight to each element, which reflects the repercussion of each of them, all the elements are computed with a function resulting in attack surface level. The main advantage of this method is that dividing the metric into small elements helps to simplify the approach. However, the function for computing all the elements together have to be specified for each system, which, even for a simple system, can be extremely difficult. Additionally, this work relies on the history of attacks on a system, which prevents it for applying in a systematic form.

Continuing with attack surface, Manadhata and Wing [19] modified the attack surface metric by categorising the system resources into different *attack classes*. The main idea behind classifying system resources is based on the notion that some of them are more likely to be attacked than others. After identifying and classifying all attackable system resources, they presented, measured and compared the attack surface of two Linux distributions, two IMAP servers [20, 21, 23] and two FTP Daemons [18, 20, 23].

In the case of multiple metrics applied in ESs, Garitano et al. [6] present a methodology which evaluates the entire system SPD level. The presented methodology starts by evaluating each component of the system to jump over sub-system evaluation and end up with the entire system SPD level. However, their work is mainly focused on privacy, which will be further explained in this work.

Previous research analysed SPD elements individually. Even if some of them combine security and privacy or security and dependability, none of them combines all three (security, privacy, dependability) of them. Considering each SPD independently in designing fragile systems could result in a highly secure system, but the system might be highly vulnerable due to dependabilities. In most cases a specific SPD level requires a compromise between one or the other, implying a balance between all three of them. The next section describes a methodology to combine and evaluate different metrics by the usage of the Multi-Metrics approach.

4 Smart Grid Use Case

This section describes the evaluated Smart Grid installation in the south of Norway and analyses use cases. The applicability of the presented methodology is further explained in Section 7.

The presented Smart Grid is composed of *(i)* customer AMSs and information collector infrastructure, *(ii)* cloud services and *(iii)* remote access for monitoring and control. As shown in Figure 2, the information acquired from AMS devices is first collected by the control centre and later transmitted to the cloud services for its analysis and storage. At the same time, end customers can access, check and control remotely their home status by the cloud services.

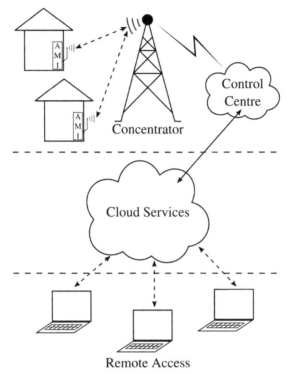

Figure 2   Smart Grid topology including radio communication, cloud services and remote access

The Smart Grid provides several use cases such as grid monitoring, different types of alarms (fire, flood), billing, home control and smart energy generation. This work analyses and evaluates three of them, *(i)* billing, *(ii)* fire alarm and *(iii)* home control.

4.1 Billing Use Case

In case of the billing use case, the meter reader provides hourly meter values and real time energy related alarms (sags, swells and power faults). This information is first sent to the concentrator and kept there until it is sent to the control centre. The concentrator provides the control centre with the meter values typically every sixth hours, while alarms are sent in real time. All communication from the concentrator to the control centre is sent over mobile network. The data are stored in a cloud infrastructure, where they are validated, missing values estimated and billing is prepared.

4.2 Alarm Use Case

Alarm services are foreseen, including e.g. a fire alarm, which is communicated directly to the fire brigade, and to the control centre. In addition, the home owner might be informed by an SMS depending on his/her profile settings.

The current infrastructure first collects the alarm in the concentrator, and then directly forwards it to the control centre. One part of the SPD analysis will address the challenges of having just one communication channel.

4.3 Home Control Use Case

The home control use case uses a bidirectional communication both with the control centre and the cloud. In case of the meter reading, it checks the status of different home devices and it sends this information to the cloud. In the current centralised version the cloud runs load control algorithms, and might send control signals to the home. Future installations might address a local processing in the AMI infrastructure in the home. Furthermore, customers will gain the opportunity to check and change the settings that will be transmitted back to the AMI.

The presented Smart Grid has three main parts and multiple use cases. However, the analysis and SPD evaluation of all them is not feasible within this work. Thus, in Section 7, the SPD level for the customer AMSs and information collector infrastructure will be evaluated for the three use cases presented above. The evaluation will provide a clear clue whether the system will run according to the established SPD_{Goal}s or whether it will be necessary to redesign it to accomplish the set SPD requirements.

5 Methodology for Security, Privacy and Dependability

This section describes the methodology together with the system structure followed to evaluate the system Security, Privacy and Dependability, SPD_{System}, level. The main objective is to evaluate multiple system configurations and select those which address or achieve the established requirements.

5.1 System SPD Evaluation

The SPD_{System} level is represented by a triplet composed of individual Security, Privacy and Dependability levels (s, p, d). Furthermore, each element is described by a value in a range between 0 and 100, i.e. the higher the number, the higher the Security, Privacy and Dependability levels.

The given methodology concludes with the SPD$_{System}$ level, however, the system criticality is used during the whole process as the main evaluation component. As such, as well as SPD$_{System}$, criticality is a triplet defined as the complement of SPD, and expressed as $(Cs, Cp, Cd) = (100, 100, 100) - (s, p, d)$.

A single system could work under different use cases, e.g. fire alarm, consumption measurement and billing in the case of the Smart Grid. Furthermore, depending on the use case, the system can be configured in multiple ways. Thus, for a given use case and system, there will be different configuration options. Besides multiple configurations, each system use case has a required SPD level, SPD$_{Goal}$. Moreover, each configuration offers a different SPD level, hence, the proposed methodology evaluates all possible configurations looking for the most convenient one.

The evaluation process, for a given system configuration, starts by evaluating each single component to end up with the whole system evaluation. The scalability allowing an analysis of individual sub-systems simplifies the process complexity; at the same time it helps to identify the main risk sources. As shown in Figure 3, a system is composed of multiple sub-systems which at the same time consist of various components. The evaluation of components and sub-systems is performed by multiple metrics and the Multi-Metrics process, further explained in Section 7.

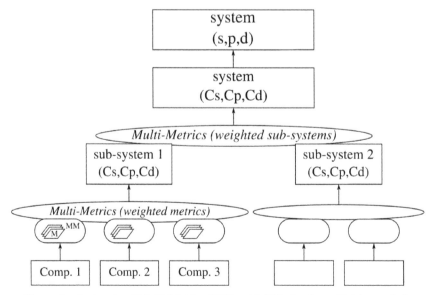

Figure 3   System level Multi-Metrics (MM), with M indicating a Metrics analysis

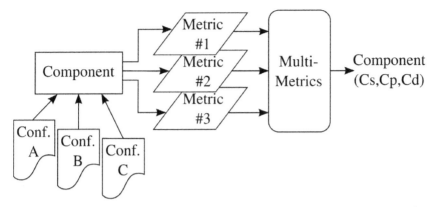

Figure 4   Component level Multi-Metrics (MM), with M indicating a Metrics analysis.

Metrics are objects or entities used to measure the criticality of components. As presented in Figure 4, the criticality of a component for a given configuration is evaluated through one or more metrics. The result of the metrics is again joined by using Multi-Metrics, which provides the overall component measurement. The selection and/or definition of metrics, applied for a specific use case, is further explained in Section 6.

The SPD_{System} evaluation is performed individually for each SPD component. The last step is to make a commitment and select the most convenient configuration for a given scenario. To check the SPD_{System} compliance, a visual representation is used. The main advantage is the simplification of the comparison between the SPD level of each configuration and the established SPD_{Goal}. Thus, every element of SPD level is substituted by a *green, yellow* or *red* circle. The colour is selected according to the numeric difference between SPD level and SPD_{Goal}, following the following criteria:

- $|SPD_{Goal}-SPD\ level| = < 10$, green ⓖ.
- $|SPD_{Goal}-SPD\ level| = > 10, < 20$, yellow ⓨ.
- $|SPD_{Goal}-SPD\ level| = > 20$, red ⓡ.

As result, the selection of the most convenient configuration will establish the SPD_{System} level.

5.2 Weight of the Elements

As shown in Figure 3, a system is composed of components and sub-systems. Additionally, the evaluation of component criticality can be performed by one or multiple metrics, as shown in Figure 4.

The importance of each component, metric or sub-system is not equal for the system operation, as is different for SPD$_{System}$ evaluation. As such, all the elements involved in the system evaluation have a weight. The weight describes the importance of a configuration for a sub-system, e.g. authentication or remote access, and is further described in Section 7.

As well as SPD levels, the weight is described by a value in a range between 0 and 100, i.e. the higher the number, the higher the importance of an element. The usage of the same range for SPD and element importance helps to assign the corresponding value. However, in order to increase the weight effect in the Multi-Metrics, the initial weight value is escalated quadratically. Our sensitivity analysis showed that a linear weighting results in an averaging effect for the SPD value of sub-systems and systems, rather than pin-pointing criticalities. The Multi-Metrics approach is further explained in Section 7.

5.3 Use Case Based Goal

The last step of the methodology is to compare the SPD level of each configuration with the SPD$_{Goal}$ established for each use case. Thus, one of the first requirements is to set a SPD$_{Goal}$ for each use case. If the SPD level of the closest system configuration to SPD$_{Goal}$ is still far from the objective, the result will show the need for a new configuration or a redesign of the system.

Table 1 shows the established SPD$_{Goal}$ values for the three use cases analysed in this work. As it is shown, *Billing* and *Home Control* are focused first on security leaving dependability as the less important one. This is mainly due to the fact that in both cases is necessary to avoid any kind of man in the middle attack. Furthermore, the response time of the system does not have to be immediate, thus, decreasing the dependability level. However, in case of *Alarm*, dependability is the principal target, then security and finally privacy. A fire alarm needs a fast delivery of the alarm with a high reliability, while protecting the privacy of the communication is of less importance.

In this section, the methodology for evaluating and selecting the best SPD level according to the SPD$_{Goal}$ has been presented. Furthermore, we introduced how different elements are weighted and defined the SPD$_{Goal}$ for

Table 1   SPD$_{Goal}$ of each use case

Use Case	Security	Privacy	Dependability	SPD$_{Goal}$
Billing	90	80	40	(90,80,40)
Home Control	90	80	60	(90,80,60)
Alarm	60	40	80	(60,40,80)

three specific use cases. The main advantage of the methodology consists of the simplicity of evaluating and selecting the most appropriate configuration for a given use case. The Multi-Metrics approach reduces the complexity of the evaluation process while the visual representation of SPD simplifies the selection process.

6 Sub-System and Metrics

This section describes the three sub-systems which compose the presented system and the selection and definition of SPD metrics. Six metrics are used to evaluate the SPD_{System} level of the Smart Grid presented in Section 4.

6.1 Sub-Systems Description

The Smart Grid system presented in this paper is composed of multiple sub-systems and components. This section describes three of them, *(i)* the Automatic Meter Reader (AMR), *(ii)* the Mesh radio link and *(iii)* the Mobile link sub-systems. These sub-systems are used by the three use cases described in the previous section.

As shown in Figure 1, the AMR uses the Mesh radio link to communicate with the signal concentrator. The concentrator uses the Mobile link to send and receive data from the Control Centre. Thus, all three sub-systems communicate with each other. Given the fact that different sub-systems are interrelated, the whole system could be evaluated together. However, the analysis of the overall SPD_{System} would became complex thus, each subsystem is evaluated individually. The division of the system into sub-systems and each sub-system into several components, allows the easy identification and evaluation of the metrics.

The AMR is an Embedded System (ES) installed in every house, and is tailored to measure, sense and in the future control the power consumption, fire sensors and some other home parameters. While current AMR are monitoring, they will be extended allowing end-users to control the home through the operator's infrastructure.

The Mesh radio link is the communication channel used by the AMR and the concentrator to communicate with each other. Since the installation foresee one concentrator per a group of houses, the communication between each AMR and the concentrator can be done directly or by multiple hops. In case of direct communication the transmit power for the direct communication from each house to the concentrator will be higher as compared to a mesh set-up, increasing the chance of interference. In a non-synchronised wireless network, communications from each AMR to the concentrator may also increase the probability of signal collisions. In a mesh configuration, data can be transmitted by using multiple hops. In this case the probability of reaching the concentrator are bigger, since data can follow multiple routes.

The communication between the concentrator and the Control Centre is performed by the Mobile link sub-system. Being controlled by the service provider, the mobile communication system can choose between sending the data over SMS or GPRS.

6.2 Sub-Systems Metrics

The SPD level of the components that make up a sub-system can be measured by multiple metrics. Definition and selection of the necessary metrics requires expertise in the field, and should be performed by a system engineer. One of the ideas behind this work resides in the creation and maintenance of a common metric database. The main benefit consists of reusing the metrics used to measure the same or equivalent component SPD level, or even use existing SPD values for sub-systems/components with a given configuration. To the best of our knowledge, there is no metric database available, giving us the task of defining relevant metrics.

The definition of a metric starts by analysing every component and identifying all parameters which could be used for its characterization. Those parameters have to be evaluated from all SPD perspectives in order to end up with a criticality level for each of them. Next step is to evaluate the repercussion of the possible values of each parameter on the component SPD level. The output will be the weight, which could vary from one system to another and thus, needs to be defined or at least evaluated for every new system evaluation.

In case of the presented three sub-systems each of them is evaluated by two or three metrics, having a total of six metrics. The evaluation of the AMI sub-system, is performed by *(i)* Remote Access, *(ii)* Authentication, and *(iii)* Encryption metrics.

The **Remote Access** metric evaluates the SPD level of the remote connectivity functionality of the system. As shown in Table 2, this metric establishes different criticality values for whenever the functionality is activated or not.

Authentication metric, Table 3, establishes the criticality level of having authentication activated in order to access the AMI. It considers both, remote as well as local access to the AMI.

Table 2 Remote access metric

Configuration	Cs	Cp	Cd
Remote Access ON	60	60	40
Remote Access OFF	10	20	50

Table 3 Authentication metric

Configuration	Cs	Cp	Cd
Authentication ON	10	30	60
Authentication OFF	80	70	40

The third metric used to measure the SPD level of the AMI is **Encryption**. This metric is used in all three sub-systems to evaluate if the transmitted data is encrypted or not. As shown in Table 4, it considers two different status, data encryption activated or not.

The evaluation of the Mesh radio link sub-system is performed by *(i)* Mesh, *(ii)* Message Rate, and *(iii)* Encryption metrics.

The traffic routing in a **Mesh** link can be performed by sending the data directly or not. In case of direct data delivery a single hop is used, which is more secure and privacy aware since data is not going through others but, requires more transmission power and the dependability is not as high. Alternatively, multiple hops traffic routing is used whenever transmission time is not as urgent and there is a need to avoid collisions. Furthermore, the necessary transmission power is lower than single hop and is more dependable since multiple paths can be used to deliver data. Table 5 shows the criticality values for Mesh metric.

Message Rate metric measures the criticality level according to the frequency the messages are sent, see Table 6. In this way, more messages per unit of time increases security and privacy criticality and reduces the dependability criticality.

Table 4 Encryption metric

Configuration	Cs	Cp	Cd
Encryption ON	10	10	60
Encryption OFF	80	80	40

Table 5 Mesh metric

Configuration	Cs	Cp	Cd
Multi-path routing	60	60	30
Single-path routing	30	30	50

Table 6 Message rate metric

Configuration	Cs	Cp	Cd
1 hour	20	20	70
20 min	25	30	50
1 min	40	50	30
5 sec	50	70	10

Table 7 Mobile channel metric

Configuration	Cs	Cp	Cd
GPRS	60	70	70
SMS	40	50	20

As well as in AMI sub-system, Encryption metric is used in Mesh radio link sub-system to evaluate if the transmitted data is encrypted or not. Same criticality values are used, shown in Table 4.

The evaluation of the Mobile link sub-system is performed by *(i)* Mobile Channel and *(ii)* Encryption metrics.

Since the mobile link is under the service provider control, the system can choose over which communication type will send data. Thus, the **Mobile Channel** metric establishes the criticality level of sending data over SMS or GPRS, as described in Table 7.

As previously explained, the Encryption metric is also used to evaluate the criticality of the Mobile link sub-system. The difference in its evaluation for each sub-system will be established by the weight.

This section introduces the three sub-systems which compose the evaluated system together with the metrics used for components criticality evaluation. Furthermore, the six metrics and their criticality values are presented. The result of evaluating the three sub-systems is further explained in the following section.

7 Multi-Metrics Approach and Operation

This chapter describes the Multi-Metrics (MM) approach and shows its applicability by analysing three different smart grid use cases.

Multi-Metrics is the core process of the overall methodology. It is a simple method which evaluates the repercussion of each metric, component or subsystem, based on its importance in the system.

During the SPD evaluation of an entire system, Multi-Metrics is used in repeated occasions to evaluate the SPD level step by step and end up with the overall SPD$_{system}$. Furthermore, the usage of the same operator along the whole SPD$_{system}$ evaluation simplifies the methodology by making it more understandable for people not being experts in the field. The output of Multi-Metrics is a single number which shows the criticality level of the components and sub-systems, and is easily translated into an SPD level.

The Multi-Metrics approach is based on two parameters: the actual criticality *xi* and the weight W_i. The criticality C is accomplished by the Root Mean Square Weighted Data (RMSWD) formula shown in Equation 1.

$$C = \sqrt{\sum_i \left(\frac{x_i^2 W_i}{\sum_i^n W_i} \right)} \tag{1}$$

There are three possible criticality level outcomes, being *(i)* component criticality, after evaluating the suitable metrics, *(ii)* sub-system criticality, from the evaluation of components or *(iii)* system criticality, after performing the Multi-Metrics operation on sub-systems. The actual criticality x_i is the result of *(i)* the metric for a component evaluation, *(ii)* the component evaluation, obtained by a previous RMSWD, for a sub-system evaluation, or *(iii)* the sub-system evaluation, obtained by a previous RMSWD, for a System evaluation. All these values are for a given configuration in a specific use case.

The weight w_i is provided by the expert in the field, and provides the significance level of each *(i)* metric within a component, *(ii)* component within a sub-system or *(iii)* sub-system within the system evaluation. As already mentioned in Section 5, the weight value is in the range of 0 to 100. Thus, it follows the same approach as the criticality level, making the entire process under the same logic. However, a sensitivity analysis has shown that a linear significance level of the weight is not appropriate to end up with representative SPD levels. Hence, the weight used in the RMSWD calculation of Equation 1 is W_i, being calculated from w_i through Equation 2 as

$$W_i = \left(\frac{w_i}{100} \right)^2 \tag{2}$$

The resulting value will be in the range of 10^{-4} and 1, maximizing the impact of high weight values towards the lowest ones.

7.1 Multi-Metrics for Smart Grid Evaluation

This section performs the evaluation of the sub-systems and components and thus demonstrates the applicability of the methodology.

The Smart Grid analysed in this work is composed by three different subsystems. As shown in Table 8, each sub-system receives a specific weight for its evaluation. Furthermore, each sub-system is composed by multiple components and in the same way, each component has a specific weight. Later, those assigned values are transformed before they are used by the RMSWD formula.

The Smart Grid evaluated in this work can run under 11 possible configurations. Those configurations set how different sub-systems and, more specifically each component, behave under some given conditions. In order to choose, for each use case, the most suitable configuration, all configurations are evaluated and compared with the established use case SPD_{Goal}.

As shown in Table 9, from 11 possible configurations the closest to the SPD_{Goal} is selected for each use case. In case of the Billing and Home Control use cases, the selected configuration is conf. 10. Both use cases are focused on security and privacy, letting dependability be of minor importance. Thus, the selected configuration, even if in both cases the security level is in red, is the most closest to security and privacy goals.

The Alarm use case is focused on dependability. Thus, the selected configuration is the one with the highest dependability value, even if it the difference is bigger than ten units. Hence, it appears in yellow.

This chapter explained the Multi-Metrics approach and showed its applicability using three Smart Grid use cases as example. As it has been shown, in order to end up with a specific configuration which best satisfies the SPD_{Goal}

Table 8   Sub-systems and components weights

Sub-system	Sub-sys. Weight	Component	Comp. Weight
AMS	80	Remote Access	70
		Authentication	80
		Encryption	80
Radio link	50	Mesh	60
		Message Rate	80
		Encryption	40
Mobile link	20	Mobile link	70
		Encryption	40

Table 9   Selected configuration SPD level for each use case

Use case	SPD_{Goal}	Configuration	SPD level	SPD vs SPD_{Goal}
Billing	(90,80,40)	10	(67,61,47)	(R,Y,G)
Home Control	(90,80,60)	10	(67,61,47)	(R,Y,Y)
Alarm	(60,40,80)	6	(31,33,63)	(R,Y,Y)

of each use case, it is necessary to set which SPD element is the major one and consider the rest as complementaries. The final result, being the SPD_{System}, is a triplet with the measured security, privacy and dependability values obtained from the application of a given system configuration.

8 Evaluation

This section evaluates the applicability of Multi-Metrics approach and the presented methodology.

The presented methodology considers all SPD aspects during the analysis of the most suitable configuration for each use case. The obtained results show under which SPD conditions the system will run for a given use case and configuration. During the design phase, the presented methodology provides a clear view of which configurations are suitable to run the system in the expected conditions or which other measures are needed to improve the SPD_{System} level. The results clarify if it is necessary to modify the design of some system aspects, and to satisfy the established goals.

The outcome of the system analysis, presented in the previous Section, showed the closest configuration options for the established SPD_{Goal}s. In all use cases the most highest SPD element is the one which dominates the selection of the configuration. Hence, in case of Billing and Home Control, the security element is the decisive one, while dependability is the decisive element for the Alarm case.

As it is shown, for Billing and Home Control the obtained security result is in red, indicating that the current configuration is not suitable for satisfying the security goals or that the security goal of $S = 90$ is unrealistic for the envisaged use case. The analysis presented configuration 10 as the one with the maximum security value of $S = 67$. A detailed sensitivity analysis shows that a specific configuration focussing just on security measures could provide a maximum value of $S = 84$. This value can satisfy the security requirements for both use cases. Furthermore, the same configuration would produce a privacy level of $P = 77$, and a dependability level of $D = 42$. These results will provide a perfect match in case of Billing $^{(G, G, G)}$ use case and a good result in case of the Home Control $^{(G, G, Y)}$ use case.

Following the same example, the results obtained for the Alarm use case show dependability and privacy in yellow and security in red. However, for this use case, even if a configuration totally focused on dependability would be created, the maximum value would be $D = 64$. Hence, in order to increase the dependability, the system needs to be redesigned by adding some other dependability focussed components, without decreasing the security and privacy values.

In case of an existing system, the same analysis will provide a clear picture about the SPD$_{System}$ level in operation. This analysis will identify which configuration options or system parts are not behaving as expected, thus help to identify the critical sub-systems. The early correction of misbehaving configuration options could prevent further consequences.

The applicability of the presented methodology is determined by the subjective weighting and criticality assignment. There is a need for a standardization through industrial interest board in order to establish the metrics, their criticality levels and their weight in a system. Without a common understanding, it is less likely that applying the methodology to a different system will yield comparable SPD levels.

However, our analysis shows that the Multi-Metrics methodology can be used to compare the SPD aspects for a given system under different configurations both during the design process or for an already existing system.

The methodology adoption from the system developers side can bring several advantages such as already evaluated metrics, components and sub-systems for different use cases. This would dramatically simplify the evaluation process and would drive its adoption from the whole industry.

9 Conclusion

This paper presents a methodology for assessing security, privacy and dependability (SPD) of embedded systems. Embedded Systems evolved from isolated to highly interconnected devices, becoming the key elements of the Internet of Things. Our approach combines the assessment of SPD, thus allows the optimisation towards a balanced solution.

In order to address the challenge of a balanced solution, the Multi-Metrics methodology presented in this paper considers all SPD aspects together. The methodology, developed through the European collaboration SHIELD, is applied for the smart grid network as deployed in the South of Norway. Three use cases, billing, home control, and alarm, are analysed in detail.

The main advantages of the methodology are the simplicity, Multi-Metrics is the core process used along all the steps, and scalability, it starts with component evaluation to jump over sub-systems and ends up with the entire system evaluation. The result is an overall SPD$_{System}$ level, which makes it easy to understand under which configuration the system will perform as envisaged by the SPD$_{Goal}$. This SPD$_{Goal}$ is defined for each of the three use cases, and the comparison with the SPD$_{System}$ shows that the system configuration can not always satisfy the envisaged goal. The paper analyses a total of 11 configurations, and concentrates on the radio communication from the meter to the control centre. As an example, assuming that billing has a security goal of 90, our analysis shows that even the configuration with the highest security settings would only reach a system security of $S = 67$.

A detailed sensitivity analysis provided an alternative configuration being able to achieve *(S, P, D)* = (84, 77, 42), satisfying the need of the billing and alarm use case. The analysis pointed out further that a single configuration is not sufficient to satisfy the given goals for all three use cases.

However, the methodology identifies solutions that are capable of satisfying security, privacy and dependability in a combined matter, and introduces measurable security for embedded systems.

Acknowledgements

The authors would like to thank their colleagues from the ARTEMIS project nSHIELD for the basics of the methodology, and the ongoing discussions on applicability. The work became part of the national ASSET initiative for security in the Internet of Things, and is financed in part by the JU ECSEL and the Research Council of Norway.

References

[1] Sarfraz Alam, Mohammad M. R. Chowdhury, and Josef Noll. Interoperability of Security-Enabled Internet of Things. *Wireless Personal Communications*, 61(3):567–586, 2011.

[2] Alexandre Bartel, Jacques Klein, Yves Le Traon, and Martin Monperrus. Automatically securing permission-based software by reducing the attack surface: an application to android. In *Proceedings of the 27th IEEE/ACM International Conference on Automated Software Engineering*, pages 274–277. ACM, 2012.

[3] Nasim Beigi Mohammadi, Jelena Mišić, Vojislav B Mišić, and Hamzeh Khazaei. A framework for intrusion detection system in advanced metering infrastructure. *Security and Communication Networks*, 7(1):195–205, 2014.

[4] Ann Cavoukian, Jules Polonetsky, and Christopher Wolf. Smartprivacy for the smart grid: embedding privacy into the design of electricity conservation. *Identity in the Information Society*, 3(2):275–294, 2010.

[5] Min Chen, Shiwen Mao, and Yunhao Liu. Big data: A survey. *Mobile Networks and Applications*, 19(2):171–209, 2014.

[6] Iñaki Garitano, Seraj Fayyad, and Josef Noll. Multi-Metrics Approach for Security, Privacy and Dependability in Embedded Systems. *Wireless Personal Communications*, accepted for publication, 2015.

[7] V.C. Gungor, Bin Lu, and G.P. Hancke. Opportunities and challenges of wireless sensor networks in smart grid. *Industrial Electronics, IEEE Transactions on*, 57(10):3557–3564, Oct 2010.

[8] V.C. Gungor, D. Sahin, T. Kocak, S. Ergut, C. Buccella, C. Cecati, and G.P. Hancke. A survey on smart grid potential applications and communication requirements. *Industrial Informatics, IEEE Transactions on*, 9(1):28–42, Feb 2013.

[9] Michael Howard. Fending off future attacks by reducing attack surface. http://msdn.microsoft.com/en-us/library/ms972812.aspx. [Online] Accessed: 2014-09-27.

[10] Michael Howard, Jon Pincus, and Jeannette M Wing. Measuring relative attack surfaces. In D. T. Lee, S. P. Shieh, and J. D. Tygar, editors, *Computer Security in the 21st Century*, pages 109–137. Springer US, 2005.

[11] G. Kalogridis, C. Efthymiou, S.Z. Denic, T.A. Lewis, and R. Cepeda. Privacy for smart meters: Towards undetectable appliance load signatures. In *Smart Grid Communications (SmartGridComm), 2010 First IEEE International Conference on*, pages 232–237, Oct 2010.

[12] Stamatis Karnouskos, Orestis Terzidis, and Panagiotis Karnouskos. An advanced metering infrastructure for future energy networks. In *New Technologies, Mobility and Security*, pages 597–606. Springer, 2007.

[13] Anil Kurmus, Alessandro Sorniotti, and Rüdiger Kapitza. Attack surface reduction for commodity os kernels: trimmed garden plants may attract less bugs. In Proceedings of the Fourth European Workshop on System Security, page 6. ACM, 2011.

[14] M.G. Lauby. Reliability considerations for application of smart grid technologies. In *Power and Energy Society General Meeting, 2010 IEEE*, pages 1–4, July 2010.

[15] C. Laughman, Kwangduk Lee, R. Cox, S. Shaw, S. Leeb, L. Norford, and P. Armstrong. *Power signature analysis. Power and Energy Magazine, IEEE*, 1(2):56–63, Mar 2003.

[16] Husheng Li and Weiyi Zhang. Qos routing in smart grid. In *Global Telecommunications Conference (GLOBECOM 2010), 2010 IEEE*, pages 1–6, Dec 2010.

[17] Zhuo Lu, Xiang Lu, Wenye Wang, and C. Wang. Review and evaluation of security threats on the communication networks in the smart grid. In *MILITARY COMMUNICATIONS CONFERENCE, 2010 - MILCOM 2010,* pages 1830–1835, Oct 2010.

[18] Pratyusa Manadhata, Jeannette Wing, Mark Flynn, and Miles McQueen. Measuring the attack surfaces of two ftp daemons. In *Proceedings of the 2nd ACM workshop on Quality of protection,* pages 3–10. ACM, 2006.

[19] Pratyusa Manadhata and Jeannette M. Wing. Measuring a system's attack surface. Technical report, DTIC Document, 2004.

[20] Pratyusa K. Manadhata, Kymie M. Tan, Roy A. Maxion, and Jeannette M. Wing. An approach to measuring a system's attack surface. Technical report, DTIC Document, 2007.

[21] Pratyusa K. Manadhata and Jeannette M. Wing. An attack surface metric. Technical report, DTIC Document, 2005.

[22] Pratyusa K Manadhata and Jeannette M Wing. An attack surface metric. *Software Engineering, IEEE Transactions on,* 37(3):371–386, 2011.

[23] Pratyusa K. Manadhata and Jeannette M. Wing. A formal model for a system's attack surface. In *Moving Target Defense,* volume 54, chapter Creating Asymmetric Uncertainty for Cyber Threats, pages 1–28. Springer New York, 2011.

[24] Yilin Mo, T. H. -H. Kim, K. Brancik, D. Dickinson, Heejo Lee, A. Perrig, and B. Sinopoli. Cyber-physical security of a smart grid infrastructure. *Proceedings of the IEEE,* 100(1):195–209, Jan 2012.

[25] nSHIELD. New embedded Systems arcHItecturE for multi-Layer Dependable solutions. http://www.newshield.eu. [Online] Accessed: 2014-09-30.

[26] Elias Leake Quinn. Privacy and the new energy infrastructure. *Available at SSRN 1370731,* 2009.

[27] Nico Saputro and Kemal Akkaya. On preserving user privacy in smart grid advanced metering infrastructure applications. *Security and Communication Networks,* 7(1):206–220, 2014.

[28] Jeffrey Stuckman and James Purtilo. Comparing and applying attack surface metrics. In *Proceedings of the 4th international workshop on Security measurements and metrics,* pages 3–6. ACM, 2012.

[29] Jakub Szefer, Eric Keller, Ruby B Lee, and Jennifer Rexford. Eliminating the hypervisor attack surface for a more secure cloud. *In Proceedings of the 18th ACM conference on Computer and communications security,* pages 401–412. ACM, 2011.

[30] Jeffrey Voas, Anup Ghosh, Gary McGraw, FACF Charron, and Keith W Miller. Defning an adaptive software security metric from a dynamic software failure tolerance measure. In *Computer Assurance, 1996. COMPASS'96, Systems Integrity. Software Safety. Process Security. Proceedings of the Eleventh Annual Conference on,* pages 250–263. IEEE, 1996.

[31] Jeffrey Voas and Keith W Miller. Predicting software's minimum-time-to-hazard and mean-time-to-hazard for rare input events. In *Software Reliability Engineering, 1995. Proceedings., Sixth International Symposium on,* pages 229–238. IEEE, 1995.

[32] Wenye Wang and Zhuo Lu. Cyber security in the smart grid: Survey and challenges. *Computer Networks,* 57(5):1344–1371, 2013.

[33] I.A. Whyte. Distribution network powerline carrier communication system, March 2 1976. US Patent 3,942,170.

Biographies

J. Noll is professor at the University of Oslo in the area of Wireless Network and Security. His work concentrates on personalised and context-aware service provisioning, and measurable security for the Internet of Things (IoT). He is also Head of Research in Movation, Norway's open innovation company. He is founding member of the Center for Wireless Innovation, the collaboration of 7 Universities/University colleges in Norway. He is involved in several international projects, including nSHIELD for measurable security in IoT systems, Citi-Sense-MOB for mobile air quality measurements, GravidPluss for mobile diabetes advise, and Ka-band propagation for polar regions. In the area of Internet of Things he was project leader of the Artemis pSHIELD project. Previously he was Senior Advisor at Telenor R & I in the Products and Markets group, and project leader of Eurescom's 'Broadband services in the Intelligent Home' and use-case leader in the EU FP6 'Adaptive Services rid (ASG)' projects, and has initiated a.o. the EU's 6th FP ePerSpace and several Eurescom projects. In 2008 he received the IARIA fellow award. He is editorial board member of four International Journals, as well as reviewer and evaluator for several national and European projects and programs.

I. Garitano is currently working as a postdoctoral fellow at UNIK-University Graduate Centre, Norway. He received the Ph.D. degree from the Department of Electronics and Computer Science, University of Mondragon in 2014 in the area of industrial control systems security. Prior to that he received the M.Sc. degree in Telecommunication Engineering from University of Mondragon. His current research interests include measurable Security, Privacy and Dependability (SPD), Intrusion Detection Systems (IDS) and Internet of Things (IoT). He participated, and currently is involved, in research projects funded by the Norwegian Research Council, the Basque Government, the Spanish Government and the European Union.

S. Fayyad, PhD researcher at Movation AS and the University of Oslo/UNIK, he received his M.Sc. degree in computer engineering in the area of «reliable systems» from the University Duisburg-Essen, Germany. His research interests include IT security with concentration on measurable security for sensors in the Internet of People, Things and Services (IoPTS). He is involved in several international projects, including nSHIELD for measurable security in IoT systems, Citi-Sense-MOB for mobile air quality measurements.

E. Åsberg is currently Head of Development and Product Architect at eSmart Systems. He received his degree in Software Design from Østfold University College. He started his career at Institute for Energy Technology as a systems developer, continuing at Hand-El Scandinavia working with Customer Information Systems and at Nasdaq OMX working with Risk Management Systems. At Navita Systems (later Brady Plc.) he continued as team lead for software development on their risk management system for mitigating risk in the financial energy and commodity markets. At eSmart Systems he is responsible for system development and overall system architecture. He is heavily involved in system specification working closely with the customers. eSmart base their architecture on Azure, Microsoft's cloud solution, and Erik has extensive knowledge of the services available on the platform and works closely with Microsoft to optimize eSmart's cloud based solutions.

Dr. H. Abie is currently a senior research scientist at the Norwegian Computing Center. He received his B.Sc., M.Sc. and Ph.D. from the University of Oslo. He has previously been a scientific associate and fellow at CERN, researcher at ABB Corporate Research, Norway, software development engineer at Nera-AS, Norway, Alcatel Telecom Norway AS, Oslo, Norway, and senior engineer and research scientist at Telenor R & D, Norway. He has a solid and extensive background in the design and development of real-time systems, and the design, modelling and development of security for critical systems. He participates as a reviewer and member of the technical program committee in international conferences and workshops and reviews scientific papers in books and international journals. He co-organizes international workshops in conjunction with highly reputed international conferences, and serves as a project proposal reviewer for research and higher academic institutions. His past and present research interests encompass adaptive security, privacy and trust in distributed and communications systems, architecture and methodology, formal methods and tools, hard real-time systems, and mobile, ubiquitous, Internet of Things (IoT), and ambient intelligent computing, and adaptive and evolving algorithms.

Study on Estimating Buffer Overflow Probabilities in High-Speed Communication Networks

Izabella Lokshina

Department of Management, Marketing and Information Systems
SUNY–Oneonta, NY 13820, USA
Izabella.Lokshina@oneonta.edu

Received January 2015; Accepted 18 February 2015;
Publication 3 April 2015

Abstract

The paper recommends new methods to estimate effectively the probabilities of buffer overflow in high-speed communication networks. The frequency of buffer overflow in queuing system is very small; therefore the overflow is defined as rare event and can be estimated using rare event simulation with continuous-time Markov chains. First, a two-node queuing system is considered and the buffer overflow at the second node is studied. Two efficient rare event simulation algorithms, based on the Importance sampling and Cross-entropy methods, are developed and applied to accelerate the buffer overflow simulation with Markov chain modeling. Then, the buffer overflow in self-similar queuing system is studied and the simulations with long-range dependent self-similar traffic source models are conducted. A new efficient simulation algorithm, based on the RESTART method with limited relative error technique, is developed and applied to accelerate the buffer overflow simulation with SSM/M/1/B modeling using different parameters of arrival processes and different buffer sizes. Numerical examples and simulation results are provided for all methods to estimate the probabilities of buffer overflow, proposed in this paper.

Journal of Cyber Security, Vol. 3, 399–426.
doi: 10.13052/jcsm2245-1439.343

Keywords: high-speed communication networks, estimating probability of buffer overflow; two-node queuing system with feedback; Importance sampling method, Cross-entropy method; self-similar queuing system; RESTART method with limited relative error technique.

1 Introduction

Overflows in high-speed communication networks are uncommon and defined as rare events. The frequency of rare events is very small, e.g. 10^{-6} or less; however, rare event probability can be used to simulate, estimate and analyze many queuing system characteristics. Queuing systems are appropriate reference models being used in different mehodologies and techniques to accelerate rare event simulation in high-speed communications networks. Estimation of rare event probability using Monte Carlo simulation requires a very long computing time and cannot easily be implemented [2, 4, 19]. Lately two basic methods of the rare event simulation were developed based on cross-entropy approach that can be applied to a wide range of optimization tasks [6, 9, 10]:

- Splitting of the sample path that for to reach definite intermediate level between the starting level and rare event [5]; and
- Importance Sampling (IS) generation [3].

The Probability Density Function (PDF) is used in the IS approach as a rare event evaluation measure, which can be compared and changed based on the likelihood ratio of the less rare event PDF [17].

One of the rare event simulation objectives is estimating total network population overflow. Exact large deviation analysis leading to asymptotically efficient change of measure is rather difficult. Instead, heuristic change of measure is proposed, which interchanges the arrival rate to the first queue and the slowest service rate. Similar change of measure is suggested based on time reversal arguments. However, analysis shows that the IS estimator based on this change of measure is not necessarily asymptotically efficient. In fact, it has an infinite variance in some parameter regions [10].

Another rare event simulation objective is estimating buffer overflow observed at individual network node. This objective is the purpose of our study. If the node of concern is a bottleneck relative to all preceding nodes, then asymptotically efficient exponential change of measure can be obtained by interchanging the arrival rate and the service rate at this particular node; and the service rates at all other nodes are kept unchanged [11–12].

However, this change of measure is not asymptotically efficient when overflow of buffer at the consequent node is considered. Effective bandwidth is used to derive heuristics for efficient feed-forward discrete-time queuing network simulation. This class of networks essentially resembles feed-forward fluid-flow networks. The analogous approach to continuous-time queuing networks has not yet been introduced even for Markov chains.

Initially, two-node queuing systems are considered in this paper; and the event of buffer overflow at the second node is studied. Discrete-Time Markov Chain (DTMC) with its regular structure is highly efficient model used for performance evaluation of the queuing system. On one hand, the states are easily arranged as a grid in DTMC (with as many dimensions as the number of queues). On the other hand, any transition in DTMC corresponds to a particular elementary event at one of the queues (e.g., arrival or service completion). These events are known as transition events, and they are defined separately from the states; i.e., there is only one transition event for a service completion at a given queue, and this particular transition event corresponds to a transition out of each state in the DTMC while this particular queue is non-empty [13, 16].

However, not all transition events are enabled in every state. For example, the service completion event of the particular queue is not possible in a state where the queue is empty.

In this paper we propose a new simulation method based on Markov Additive Continuous-Time Process (MAP) modeling. We develop and apply Importance sampling algorithm with exponential tilting of the unbiased PDF estimation to the appropriate MAP representation, which let us estimate effectively the probability of buffer overflow at the second node. Unlike several heuristic changes of measure described in the literature, the derived change of measure depends on the content of the first buffer.

When the first buffer is finite, we confirm that the proposed simulation procedure yields the estimation with a relative error that is bound independent of the buffer overflow level. This result is much stronger than the asymptotic efficiency, which cannot be observed with other known methods.

When the first buffer is infinite, we propose a natural extension of change of measure for finite buffer case. Applying the orthogonal polynomial model we obtain two types of simulation behavior. When the second buffer is a bottleneck, we confirm once more that simulation yields the estimation with a relative error that is bound independent of the buffer overflow level. However, when the first buffer is a bottleneck, the simulation results prove that the relative error is asymptotic and linearly bound to the buffer overflow level.

Finally, long-range dependent self-similar queuing systems are considered in this paper; and the event of buffer overflow is studied. We propose a new steady-state simulation method, based on SSM/M/1/B modeling.

We develop and apply RESTART with Limited Relative Error (LRE) algorithm, which let us estimate effectively the probability of buffer overflow in a long-range dependent self-similar queuing system with different parameters of arrival processes and different buffer sizes.

2 Estimating Probability of Buffer Overflow in Continuous Time Queueing Systems

Markov additive continuous-time process is a stochastic process (J_t, Z_t), where (J_t) is Markov chain with the denumerable state space, and (Z_t) has stationary and independent increments during the time intervals when (J_t) is in any given state. That is, if given J_t has not changed in the interval (t_1, t_2), then for any $t_1 < s_1 < \ldots < s_n < t_2$, the increments $Z_{s_2} - Z_{s_1}, \ldots, Z_{s_n} - Z_{s_{n-1}}$ are mutually independent, and the total increment during the interval $[t_1, t_2]$ depends on t_1 and t_2 only through the difference $t_2 - t_1$.

Moreover, the transition from state i to state j (J_t) has a certain probability (depending only on i and j) of triggering the transition of (Z_t) at the same time. The size of the transition in the process (Z_t) has fixed distribution, which depends only on i and j.

Markov additive continuous-time process (J_t, Z_t) is characterized by the family of the matrices $(M_t(s), t \geq 0)$, where (i, j)-th element of $M_t(s)$ is (1),

$$[M_t(s)]_{ij} = E_i[e^{s(Z_t - Z_0)} I_{\{J_t = j\}}] \tag{1}$$

where E_i, denote the expectation operator given to the initial MAP state $J_0 = i$. Let us notice that $M_t(\cdot)$ is a generalization of the moment generating function for the ordinary random variables, as shown in (2).

$$E_i[e^{s(Z_{t+h} - Z_0)} I_{\{J_{t+h} = j\}}] =$$
$$= \sum_k E_i[e^{s(Z_{t+h} - Z_0)} I_{\{J_t = k\}} I_{\{J_{t+h} = j\}}]$$
$$= \sum_k E_i[e^{s(Z_t - Z_0)} I_{\{J_t = k\}}] E_i[e^{s(Z_{t+h} - Z_t)} I_{\{J_{t+j} = j\}} | J_t = k] \tag{2}$$
$$= \sum_k [M_t(s)]_{ik} E_k[e^{s(Z_h - Z_0)} I_{\{J_h = j\}}]$$

Consequently, if for all k and j (3) can be defined,

$$[A(s)]_{kj} := \lim_{h \downarrow 0} \frac{1}{h} E_k [e^{s(Z_h - Z_0)} I_{\{J_h = j\}} - \delta_{kj}] \tag{3}$$

where δ is usual notation of Dirac, then (4) can be easily obtained,

$$\frac{d}{dt} M_t(s) = M_t(s) A(s), \qquad t \geq 0 \tag{4}$$

with $M_0(s) = I$ (the identity matrix). It is true as soon as (5) is true.

$$M_t(s) = e^{tA(s)}, \quad t \geq 0 \tag{5}$$

The matrix $A(s)$ is known as the MAP (infinitesimal) generator.

Let us consider a simple Markov chain that consists of two queues in tandem. The calls arrive to the first queue (e.g., buffer) according to Poisson process with the rate λ. The departure time is exponentially distributed with the rate μ_1. The calls that leave the first queue enter the second queue. The departure time has an exponential distribution with the rate μ_2.

The queuing system stability is assumed, i.e. $\lambda < \min \{\mu_1, \mu_2\}$. The size of the first buffer may be finite or infinite; in fact, let us consider both cases. Let X_t and Y_t denote the number of the calls in the first and the second queues at the time t, respectively. Let P_i denote the probability measure under which (X_t) starts from i at the time 0 (i.e., $X_0 = i$, $i \geq 0$); and let E_i, denote the corresponding expectation operator.

Assuming that the second buffer is initially non-empty, say, $Y_0 = 1$, the probability that, starting from $(X_0, Y_0) = (i, 1)$, content of the second buffer hits some high level L∈N before hitting 0, can be estimated. This probability is noted by γ_i, and referred to it as the second buffer overflow probability, given that the initial number of calls in the first queue is i.

2.1 Rare Event Simulation with Markov Chain Modeling

Let us consider the IS approach for the rare event simulation, where the probability density function is used as the measure of rare event evaluation, which is compared and changed with the likelihood ratio of the probability density function of a less rare event. First, let us determine the rare event. Let $X = (X_1,..., X_N)$ be a random vector, which values belong to the certain state space χ. Let $\{f(\cdot, v)\}$ be the family of the probability density functions on χ, with respect to some base measure v. Here v is a real-valued parameter (vector). Then, for any measurable function H is obtained as (6).

$$E[H(X) \int H(x)f(x; v)v(dx)]$$ (6)

In many cases f is often called the Probability Mass Function (PMF), but in this paper the generic term density, or the Probability Density Function (PDF), is used. Let S be some real function on χ. The probability that S(X) is greater than some real number γ, under $F(\cdot; u)$ can be defined. Therefore, probability can be written as (7),

$$l = P_u(S(X) \geq \gamma) = E_u[I_{\{S(X) \geq \gamma\}}]$$ (7)

where $I_{\{S(X) \geq \gamma\}}$ is the indicator function. If this probability is very small, like 10^{-6} or less, then $\{S(X) \geq \gamma\}$ is a rare event. The simplest way to estimate l is to use the basic Monte Carlo simulation. Draw a random sample $X_1,..., X_N$ from f($\cdot$;u), and use (8) as the unbiased estimator of l.

$$\frac{1}{N} \sum_{i=1}^{N} I_{\{S(X_i) \geq \gamma\}}$$ (8)

However, it poses serious problems when $\{S(X) \geq \gamma\}$ is a rare event. In that case a large simulation effort is required in order to estimate l accurately. An alternative is based on the IS. Take a random sample $X_1,..., X_N$ from the IS density g on χ, and evaluate l using the unbiased estimator, called the likelihood ratio estimator (9).

$$\widehat{l} = \frac{1}{N} \sum_{i=1}^{N} I_{\{S(X_i) \geq \gamma\}} \frac{f(X_i; u)}{g(X_i)}$$ (9)

It is well known that the optimal way to estimate l is to use the change of the measure with the density (10)

$$g^*(x) = \frac{I_{\{S(X) \geq \gamma\}} f(x; u)}{l}$$ (10)

Specifically, applying this change of the measure to (9), (11) can be obtained for all i.

$$I_{\{S(X_i) \geq \gamma\}} \frac{f(X_i; u)}{g^*(X_i)} = l$$ (11)

In other words, the estimator (9) has a zero variance, and only N = 1 sample need to be produced. The obvious difficulty is, of course, that the g* depends on the unknown parameter l. Moreover, one often wishes to choose this g from the density family $\{f(\cdot, v)\}$. Now the plan is to choose the tilting parameter v, such that the distance between the densities g* and $f(\cdot, v)$ is minimal.

A particular suitable measure of the distance between two densities g and f is the Kullback-Leibler distance, which is defined in (12).

$$D(g, f) = E_g[\ln \frac{g(X)}{f(X)}]$$

$$= \int g(x) \ln g(x) v(dx) - \int g(x) \ln f(x) v(dx) \qquad (12)$$

Therefore, minimizing the Kullback-Leibler distance between g in (11) and $f(\cdot, v)$ is the same as choosing v, such that $-\int g(x) \ln f(x; v) v(dx)$ is minimized, or equally, such that $-\int g(x) \ln f(x; v) v(dx)$ is maximized. Formally, it can be written as (13).

$$\max_v D(v) = \max_v \int g(x) \ln f(x; v) v(dx) \qquad (13)$$

Applying g from (10) to (13) as the substitution, the following optimization program can be obtained as (14).

$$\max_v D(v) = \max_v \int \frac{I_{\{S(X) \geq \gamma\}} f(x; u)}{l} \ln f(x; v) v(dx)$$

$$= \max_v E_u[I_{\{S(X) \geq \gamma\}} \ln f(X; v)] \qquad (14)$$

Using the IS again, with the change of measure $f(\cdot; w)$, (14) can be re-written into (15),

$$\max_v D(v) = \max_v E_w[I_{\{S(X) \geq \gamma\}} W(X; u, w) \ln f(X; v)] \qquad (15)$$

for any tilting parameter w, where the likelihood ratio at x between $f(\cdot; u)$ and $f(\cdot; w)$ is W. This can be presented according to (16).

$$W(x; u; w) = \frac{f(x; u)}{f(x; w)} \qquad (16)$$

The basic idea of the IS method regarding rare event is to accelerate its frequency with iterative tilting the unbiased estimation of the probability density function to appropriate MAP representation.

The acceleration of conditional probability of rare event X_1 with one-parameter function $f(x)$ is shown in Figure 1. The conditional probability of rare event $P_f\{x > X_1\}$ is changing with conditional probability of $g(x) - P_g\{x > X_1\}$.

At each iteration of the IS simulation, N independent samples are generated, which distribution $g(x)$ can be evaluated with the likelihood ratio $W(X_1)$. The optimal solution for (15) can be written as (17).

$$v^* = \arg \max_v E_w[I_{\{S(X)\geq\gamma\}} W(X; u, w)\ln f(X; v)] \qquad (17)$$

It can be obtained by solving the following stochastic program, which can be considered as a stochastic counterpart of (15) and written according to (18),

$$\max_v \widehat{D}(v) = \max_v \frac{1}{N} \sum_{i=1}^{N} I_{\{S(X)\geq\gamma\}} W(X_i; u, w)\ln f(X_i; v) \qquad (18)$$

where $X_1,..., X_N$ is a random sample from $f(\cdot;w)$.

The solution for (18) can be obtained by solving the following system of equations with respect to v in (19),

$$\frac{1}{N} \sum_{i=1}^{N} I_{\{S(X_i)\geq\gamma\}} W(X_i; u, w)\nabla \ln f(X_i; v) = 0 \qquad (19)$$

where the gradient is defined regarding v.

This confirms the expectation that the differentiation operators and the function $\widehat{D}$ can be interchanged in (18) with respect to v. The advantage of

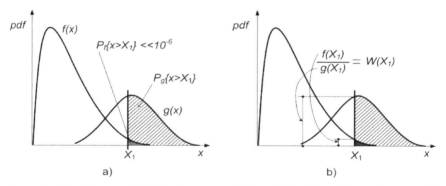

Figure 1   Conditional probability of rare event $P_f\{x > X_1\}$ - (a); and its acceleration with likelihood ratio $W(X_1)$ - (b).

this approach is in the fact that often the solution of (19) can be calculated analytically. In particular, this happens when the random variable distributions belong to the natural exponential family. The cross-entropy program (19) is useful only when the probability of the target event $\{S(X) \geq \gamma\}$ is not too small, say $1 \geq 10^{-5}$.

Then, above program is useful in terms of potentially more accurate estimator determination. However, in a rare event context, (say, $1 \geq 10^{-6}$), the program (19) is useless to rarity of events $\{S(X_i) \geq \gamma\}$, because the random variables $I_{\{S(X_i) \geq \gamma\}}$, $i = 1,..., N$ and the associated derivatives of $\widehat{D}(v)$ vanish with high probability, as given in the right-hand side of (19), for reasonable sizes of N.

2.2 Exponential Change of Measure

Let us initially think that the first buffer has a finite capacity N. In this case the state space of the driving process (X_t) is finite in $\{0,...,N\}$. Let us consider Markov additive continuous-time process (X_t, Z_t). To create a corresponding MAP generator (i.e., matrix A(s) in (5)), the infinitesimal expectations $E_i[e^{s(Z_h - Z_0)}I_{\{X_h = j\}} - \delta_{ij}]$ as $h \downarrow 0$ for all i, j in $\{0,...,N\}$ have to be determined, where $Z_0 = 1$ and $\delta_{ij} = 0$ for $j \neq i$.

For instance, since the downward transition of (X_t) leads to the upward transition of (Z_t), (20) is used for $i = 1,..., N$, as $h \downarrow 0$.

$$E_i[e^{s(Z_h - Z_0)}I_{\{X_h = i-1\}} - \delta_{i,i-1}] =$$
$$= E_i[e^{s(Z_h - Z_0)} | X_h = i - 1]P_i(X_h = i - 1) =$$
$$= e^s(\mu_1 h + O(h)) = \mu_1 h e^s + o(h) \qquad (20)$$

Therefore, the (i, i-l)-th element of the matrix A(s) exists and is equal to $\mu_1 e^s$. Other elements of the matrix A(s) can be defined similarly. Consequently, (5) holds with A(s) for Markov additive continuous-time process (X_t, Z_t) with the given $(N + 1, N + 1)$-tridiagonal matrix (21).

$$G_N(s) = \begin{pmatrix} -\lambda - \mu_2 + \mu_2 e^{-s} & \lambda & & & \\ \mu_1 e^s & -\lambda - \mu_1 - \mu_2 + \mu_2 e^{-s} & \lambda & & \\ & \ddots & \ddots & \ddots & \\ & & \mu_1 e^s & -\mu_1 - \mu_2 + \mu_2 e^{-s} \end{pmatrix}$$
$$(21)$$

Let us note that the MAP generator $G_N(s)$ is now equal to the matrix (22).

$$\hat{Q}^{(n)}(u) = \begin{pmatrix} -\lambda - \mu_2 + \mu_2 u & \lambda & & \\ \mu_1/u & -\lambda - \mu_1 - \mu_2 + \mu_2 u & \lambda & \\ & \ddots & \ddots & \ddots \\ & & \mu_1/u & -\mu_1 - \mu_2 + \mu_2 e^{-s} \end{pmatrix}$$

(22)

Next, the change of the measure based on the family of the matrices $\mathbf{G}_N(s)$ is defined. For any $s \geq 0$, $k_N(s)$: = $log(sp(M_t(s)))/t$ has to be defined, where $sp(M_t(s))$ denotes the spectral radius (or, the maximum Eigen value) of $M_t(s)$. Using (5) $k_N(s)$ can be identified as the largest positive Eigen value of $\mathbf{G}_N(s)$. Let $w(s) = \{w_k(s), 0 \leq k \leq N\}$ represent the correspondent right-eigenvector.

When the first buffer has the infinite capacity, the process (X_t, Z_t) is still Markov additive continuous-time process, but the state space for Markov process (X_t) is now infinite. Equation (5) is still used, but $A(s)$ is now given by the infinite-dimensional tri-diagonal matrix (23).

$$G(s) = \begin{pmatrix} -\lambda - \mu_2 + \mu_2 e^{-s} & \lambda & & \\ \mu_1 e^s & -\lambda - \mu_1 - \mu_2 + \mu_2 e^{-s} & \lambda & \\ & \ddots & \ddots & \ddots \end{pmatrix}$$

(23)

Let us note that the MAP generator $G(s)$ is now equal to the infinite tri-diagonal matrix (24).

$$Q(u) = \begin{pmatrix} -\lambda - \mu_2 + \mu_2 u & \lambda & & \\ \mu_1/u & -\lambda - \mu_1 - \mu_2 + \mu_2 u & \lambda & \\ & \ddots & \ddots & \ddots \end{pmatrix}$$

(24)

2.3 Importance Sampling Algorithm and Simulation Results

The Markovian network that consists of tandem queues with feedback is shown in Figure 2 and used as a simulation example, with the entry parameters follow: $\lambda_1 = \lambda_2 = 1$, $\mu_1 = \mu_2 = 6$, $p = 0.5$.

The Importance Sampling (IS) algorithm to accelerate rare event simulation with Markov chain modeling in high-speed communication networks is suggested in this paper. It consists of the following six steps provided below.

Algorithm 1 Importance sampling simulation method

Step 1. Set $t := 1$ (initialization of iteration counter). Define the likelihood ratio $v_1: = 0$ (in this case Monte Carlo simulation is appropriate).

Step 2. Generate a sample $X_1, \ldots, X_k, \ldots, X_N$, from the density $f(X_k; v_{t-1})$ in such way that for the ρ-th part of samples ($\rho = 0.01$) the condition of the rare event $S(X_k) > M$ is executed.

Step 3. Determine the full paths and sort ascending in following way $S_{(1)} \le S_{(2)} \le .. \le S(N)$.

Step 4. Calculate the conditional probability $\gamma_t = S_{[(1-\rho)N]}$.

Step 5. For each $S(X_k) > \gamma$ define the rare event indicator $I_{\{S(X_i) \ge M\}} = 1$ and then determine the likelihood ratio for the next iteration v_{t+1} according to (15), (16) and (19).

Step 6. If $\gamma_t < M$ then t: $= t+1$ and repeat steps 2, 3 and 4.

The simulation results for Poisson distribution and fixed numbers of calls n_1 and n_2, for four different overflow cases are provided in Table 1. As could be seen, the overflow probability exponentially decreases with increase of the fixed number of calls in queues n_1 and n_2. The exponential behavior depends more on the number n_1.

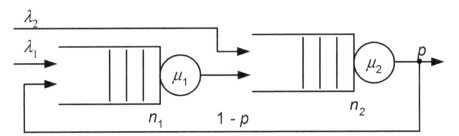

Figure 2 Two-node queuing network with feedback.

Table 1 Simulation results

Overflow Level							Overflow
n1	n2	λ_1	λ_2	μ_1	μ_2	p	Probability
25	25	0.13	0.11	0.29	0.18	0.42	2.22×10^{-8}
50	50	0.23	0.18	2.15	2.26	0.31	1.15×10^{-15}
60	40	1.18	1.42	4.91	3.58	0.44	6.29×10^{-25}
100	40	2.02	1.67	5.82	2.94	0.37	1.86×10^{-50}

2.4 Cross-Entropy Algorithm and Simulation Results

The Cross-entropy algorithm to accelerate rare event simulation with Markov chain modeling in high-speed communication networks is suggested in this paper. The idea is to introduce a sequence of reference parameters $\{v_t, t \geq 0\}$ and a sequence of the levels $\{\gamma_t, t \geq 1\}$, and iterate in both γ_t and v_t.

The initialization is done with choosing a not very small p, say $\rho = 10^{-2}$, and defining $v_0 = u$. Next, we let γ_1 ($\gamma_1 < \gamma$) be such that under the original density $f(x; u)$, the probability $1_1 = E_u[I_{\{s(x) \geq \gamma_1\}}]$ is at least ρ.

After, let v_1 be the optimal cross-entropy reference parameter to estimate l_1, and repeat the last two steps iteratively with the purpose to estimate the pair $\{l, v^*\}$. In other words, the iteration of the algorithm consists of two main phases. In the first phase γ_t is updated, in the second phase v_t is updated. Particularly, starting with $v_0 = u$, the subsequent γ_t and v_t are obtained as described below.

Phase 1 includes adaptive update of γ_t. For a fixed v_{t-1}, let γ_t be a $(1-\rho)$-quintile of $S(X)$ under v_{t-1}. That is, γ_t satisfies (25) and (26).

$$P_{v_{t-1}}(S(X) \geq \gamma_t) = E_u[I_{\{S(X) \geq \gamma\}}] \tag{25}$$

$$P_{v_{t-1}}(S(X) \leq \gamma_t) \geq 1 - \rho \tag{26}$$

where $X \sim f(\cdot \, ; \, v_{t-1})$. The simple estimator $\hat{\gamma}_t$ of γ_t can be obtained by drawing a random sample $X_1, \ldots, X_N$ from $f(\cdot \, ; \, v_{t-1})$, calculating the performances $S(X_i)$ for all i, ordering them from the smallest to the biggest: $S_{(1)} \leq \ldots \leq S_{(N)}$ and finally, evaluating the $(1 - \rho)$ sample quintile as (27).

$$\hat{\gamma}_t = S_{[(1-\rho) \, N]} \tag{27}$$

Let us note that $S_{(j)}$ is called j^{-th} order-statistic of the sequence $S(X_1), \ldots, S(X_N)$. Let us note also that $\hat{\gamma}_t$ is chosen such that the event $\{S(X) \geq \hat{\gamma}_t\}$ is not too rare (it has a probability of around ρ), and therefore, the reference parameter updated with a procedure such as (27) is not void of the meaning.

Phase 2 contains adaptive update of v_t. For fixed γ_t and v_{t-1}, let us derive v_t from the solution of the following cross-entropy program according to (28).

$$\max_v D(v) = \max_v E_{v_{t-1}}[I_{\{S(X) \geq \gamma\}} W(X; u, v_{t-1}) \ln f(X; v)] \tag{28}$$

The stochastic counterpart of (28) is shown as follows: for fixed $\hat{\gamma}_t$ and $\hat{v}_{t-1}$, derive $\hat{v}_t$, from the solution of the program according to (29).

$$\max_{v} \hat{D}(v) = \max_{v} \frac{1}{N} \sum_{i=1}^{N} I_{\{S(X_i) \geq \hat{\gamma}_t\}} W(X_i; u, \hat{v}_{t-1}) \ln f (X_i; v) \quad (29)$$

Therefore, at the first iteration starting with $\hat{v}_0 = u$, the target event is artificially made less rare with temporarily use of the level $\hat{\gamma}_1$, which is chosen smaller than γ that for to get a good estimate for $\hat{v}_1$.

The value $\hat{v}_1$ obtained in this way will make the event $\{S(X) \geq \gamma\}$ less rare in the next iteration, so the value $\hat{\gamma}_2$ can be used in the next iteration, which is closer to γ itself.

The algorithm terminates when the level is reached at some iteration t, which is at least γ and after the original value of γ can be used without getting too few samples. As mentioned before, the optimal solution of (28) and (29) can be often obtained analytically, in particular when f(x; v) belongs to the natural exponential family.

The above rationale results are placed in the multi-level Cross-entropy algorithm for accelerated rare event simulation with Markov chain modeling in high-speed communication networks. This efficient algorithm consists of the following five steps provided below.

Algorithm 2 Multi-Level Cross-Entropy Simulation Method

Step 1. Define $\hat{v}_0 = u$. Set $t = 1$. (Iteration = level counter).

Step 2. Generate a sample $X_1,...,X_N$ with the density $f(\cdot; v_{t-1})$ and compute the sample(1-ρ) quantile $\hat{\gamma}_t$ performance according to (28) with $\hat{\gamma}_t < \gamma$. Otherwise, set $\hat{\gamma}_t = \gamma$.

Step 3. Use the same sample $X_1,...,X_N$ to solve the stochastic program (29). Denote the solution by $\hat{v}_t$.

Step 4. If $\hat{\gamma} < \gamma$, then set $t = t + 1$ and reiterate from Step 2. Else, proceed with step 5.

Step 5. Estimate the rare event probability l using (30),

$$\hat{l} = \frac{1}{N} \sum_{i=1}^{N_1} I_{\{S(X_i) \geq \gamma\}} W(X_i; u, \hat{v}_T) \quad (30)$$

where T denotes the final number of iterations, or number of the levels used.

As a simulation example, let us apply this efficient algorithm to a similar tandem queue with feedback as was given in Figure 2, but at this time with the following entry parameters: $\lambda = 0.2$; $\mu_1 = 0.8$; $\mu_2 = 0.2$; $p = 0.5$.

The simulation results for generating N = 10,000 samples are shown in Table 2. As could be seen, overflow is obtained when there are j = 6 iterations. The accuracy increases up to N = 1,000,000 in the seventh iteration, and the overflow probability is obtained equal to $\hat{\ell}_{IS} = 1.342e^{-15}$.

Table 2   Simulation results

Iteration	Repetitive Trails	λ	μ_1	μ_2	p	Estimation
1	10^4	0.2	0.8	0.2	0.5	-
2	10^4	0.216	0.643	0.258	0.364	$1.426e^{-15}$
3	10^4	0.198	0.621	0.287	0.322	$1.462e^{-15}$
4	10^4	0.196	0.614	0.279	0.318	$1.436e^{-15}$
5	10^4	0.195	0.614	0.282	0.320	$1.372e^{-15}$
6	10^4	0.196	0.612	0.284	0.322	$1.318e^{-15}$
7	10^6	0.196	0.612	0.284	0.322	$1.342e^{-15}$

3 Estimating Probability of Buffer Overflow in Self-Similar Queuing Systems

Recent studies of high-speed communication network traffic have clearly shown that teletraffic (technical term, identifying all phenomena of transport and control of information within the high-speed communication networks) exhibits long-range dependent self-similar properties over a wide range of time scales. Therefore, self-similar queuing systems are appropriate reference models being also used in different methodologies and techniques to accelerate rare event simulation in high-speed communication networks [13].

Long-range dependent self-similar teletraffic is usually observed in LAN and WAN, where superposition of strictly independent alternating ON/OFF traffic models, whose ON- or OFF-periods have heavy-tailed distributions with infinite variance, can be used to model aggregate queuing network traffic that exhibits long-range dependent self-similar behavior, typical for measured LAN traffic over a wide range of time scales [14, 18].

Long-range dependent self-similar teletraffic is also observed in ATM networks: when arriving at an ATM buffer, it results in a heavy-tailed buffer occupancy distribution, and a buffer cell loss probability decreases with the buffer size not exponentially, like in traditional Markovian models, but hyperbolically [16, 18].

Furthermore, long-range dependent self-similar teletraffic is observed in the Internet as many characteristics can be modeled using heavy-tailed distributions, including the distributions of traffic times, user requests for documents, and document sizes. In IP with TCP self-similar queuing networks the transfer of files or messages shows that the reliable transmission and flow control mechanisms serve to maintain long range dependent structure included by heavy-tailed file size distributions [1].

Long-range dependent self-similar video traffic provides possibility for developing models for Variable Bit Rate (VBR) video traffic using heavy-tailed distributions [16, 18]. Therefore; we can clearly see that impact of self-similar models on the queuing and network performance is very significant.

The properties of long-range dependent self-similar teletraffic are very different from properties of traditional models based on Poisson, Markov-modulated Poisson, and related processes. More specifically, while tails of the queue length distributions in traditional teletraffic models decrease exponentially, those of long-range dependent self-similar teletraffic models decrease much slower.

Therefore, the use of traditional models in high-speed communication networks characterized by long-range dependent self-similar processes can lead to incorrect conclusions about the queuing and network performance. Traditional models can lead to over-estimation of the queuing and network performance, insufficient allocation of communication and data processing resources, and consequently difficulties in ensuring the QoS.

Self-similarity can be classified into two types: deterministic and stochastic. In the first type, deterministic self-similarity, a mathematical object is assumed to be self-similar (or fractal) if it can be decomposed into smaller copies of itself. That is, deterministic self-similarity is a property, in which the structure of the whole is contained in its parts [14, 18].

This work is focused on stochastic self-similarity. In that case, probabilistic properties of self-similar processes remain unchanged or invariant when the process is viewed at different time scales. This is in contrast to Poisson processes that lose their burstiness and flatten out when time scales are changed [18].

However, the time series of self-similar processes exhibit burstiness over a wide range of time scales. Self-similarity can statistically describe teletraffic that is bursty on many time scales [14].

One can distinguish two types of stochastic self-similarity. A continuous-time stochastic process Y_t, is strictly self-similar with a self-similarity parameter H $(1/2 < H < 1)$, if Y_{ct}, and $c^H Y_t$ (the rescaled process with time scale ct) have identical finite-dimensional probability for any positive time stretching factor c. This definition, in a sense of probability distribution, is quite different from that of the second-order self-similar process, observed at the mean, variance and autocorrelation levels [14].

The process X is asymptotically second-order self-similar with $0.5 < H < 1$, if for each k large enough $\rho_k^{(m)} \to \rho_k$, as $m \to \infty$, where $\rho_k = E[(X_i - \mu)(X_{i+k} - \mu)]/\sigma^2$.

In this work the exact or asymptotic self-similar processes are used in an interchangeable manner, which refers to the tail behavior of the autocorrelations [14, 18].

3.1 Long-Range Dependent Self-Similar Processes

We have to say that the most striking feature of some second-order self-similar processes is that the accumulative functions of the aggregated processes do not degenerate with the non-overlapping batch size m increasing to infinity. Such processes are known as Long-Range Dependent (LRD) processes [1, 14, 18].

This is in contrast to traditional processes used in modeling high-speed communication network traffic, all of which include the property that the accumulative functions of their aggregated processes degenerate as the non-overlapping batch size m increasing to infinity, i.e., $\rho_k^{(m)} \to 0$ or $\rho_k^{(m)} = 0(|k| > 0)$, for $m > 1$. The equivalent definition of long-range dependence is given as (31).

$$\sum_{k=-\infty}^{\infty} \rho_k = \infty \tag{31}$$

Another definition of LRD is presented as (32),

$$\rho_k \sim L(t)k^{-(2-2H)}, \quad \text{as } k \to \infty \tag{32}$$

where $1/2 < H < 1$ and $L(\cdot)$ slowly varies at infinity, i.e. for all $x > 0$ it could be determined as (33).

$$\lim_{t \to \infty} \frac{L(xt)}{L(t)} = 1 \tag{33}$$

The Hurst parameter H characterizes the relation in (32), which specifies the form of the tail of the accumulative function. One can show that is true for $1/2 < H < 1$, as given in (34).

$$\rho_k = \frac{1}{2}\left[(k+1)^{2H} - 2k^{2H} + (k-1)^{2H}\right] \tag{34}$$

For $0 < H < 1/2$ the process is Short-Range Dependent (SRD) and could be presented as (35).

$$\sum_{k=-\infty}^{\infty} \rho_k = 0 \tag{35}$$

For $H = 1$ all autocorrelation coefficients are equal to one, no matter how far apart in time the sequences are. This case has no practical importance in real high-speed communication network traffic modeling. If $H > 1$, then (36) is true.

$$\rho_k = \begin{cases} 1 & \text{for } k = 0 \\ \frac{1}{2}k^{2H}g(k^{-1}) & \text{for } k > 0 \end{cases} \tag{36}$$

where

$$g(x) = (1+x)^{2H} - 2 + (1-x)^{2H} \tag{37}$$

One can see that $g(x) \to \infty$ as $H > 1$. If $0 < H < 1$ and $H \neq 1/2$, then the first non-zero term in the Taylor expansion of g(x) is equal to $2H(2H - 1)x^2$. Therefore, (38) is true.

$$\rho_k/(H(2H-1)k^{2H-2}) \to 1, \quad \text{as } k \to \infty \tag{38}$$

In the frequency domain, an essentially equivalent definition of LRD for a process X with given spectral density (39),

$$f(\lambda) = \frac{\sigma^2}{2\pi} \sum_{k=-0}^{\infty} \rho_k e^{ik\lambda} \tag{39}$$

is that in the case of LRD processes, this function is required to satisfy the following property (40),

$$f(\lambda) \sim c_{f_1}\lambda^{-\gamma}, \quad \text{as } \quad \lambda \to 0 \tag{40}$$

where c_{f_1} is a positive constant and $0 < \gamma < 1, \gamma = 2H - 1 < 1$. As a result, LRD manifests itself in the spectral density that obeys a power-law in the vicinity of the origin. This implies that $f(0) = \sum_k \rho_k = \infty$. Consequently, it requires a spectral density, which tends to $+\infty$ as the frequency λ approaches 0.

For a Fractional Gaussian Noise (FGN) process, the spectral density $f(\lambda, H)$ is given by (41),

$$f(\lambda, H) = 2c_f(1 - \cos(\lambda))B(\lambda, H) \tag{41}$$

with $0 < H < 1$ and $-\pi \leq \lambda \leq \pi$, where (42) is true,

$$c_f = \sigma^2(2\pi)^{-1}\sin(\pi H)\Gamma(2H+1)$$

$$B(\lambda, H) = \sum_{k=-\infty}^{\infty} \mid 2\pi k + \lambda \mid^{-2H-1} \tag{42}$$

and $\sigma^2 = \mathrm{Var}[X_k]$ and $\Gamma(\cdot)$ is the gamma function. The spectral density $f(\lambda, H)$ in (41) complies with a power-law at the origin, as shown in (43),

$$f(\lambda, H) \to c_f \lambda^{1-2H}, \quad \text{as} \ \lambda \to 0 \tag{43}$$

where $1/2 < H < 1$.

3.2 Steady-State Simulation with SSM/M/1/B Modeling

As we have previously accepted, there is a significant difference in the queuing and network performance between traditional models of teletraffic, such as Poisson processes and Markovian processes, and those exhibiting long-range dependent self-similar behavior. More specifically, while tails of the queue length distributions in traditional models of high-speed communication network traffic decrease exponentially, those of self-similar traffic models decrease much slower (14]; [18]).

Let us consider the potential impacts of traffic characteristics, including the effects of long-range dependent self-similar behavior on queuing and network performance, protocol analysis, and network congestion controls. Steady-state simulation of long-range dependent self-similar queuing system includes:

- Generation of long-range dependent self-similar traffic [14, 18];
- Simulation of long-range dependent self-similar queuing process [14]; and
- Simulation of the overflow probability [14, 15].

This can be demonstrated with the buffer overflow simulation in SSM/M/1/B queuing systems ($B < \infty$, i.e. queuing systems with the finite buffer capacity) with long-range dependent self-similar queuing processes. In this case, the difference with M/M/1/B queuing system is that the arrival rate λ_j, into SSM/M/1/B queuing system is not a constant value. It depends on the sequential number of time-series i, the total number of observations n and the Hurst parameter H, which determine the rate of self-similarity. The analyzed SSM/M/1/B queuing system has exponential service times with constant rates $1/\mu$ as is shown in Figure 3. The flow balance equations are given below [8, 14]:

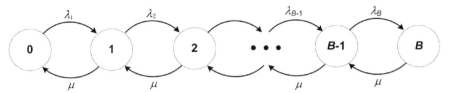

Figure 3 State transition diagram for a SSM/M/1/B self-similar queuing system.

$$\begin{aligned} \lambda_j &= \lambda(i, n, H); &\quad j &= 1, 2, \ldots, B \\ \lambda_j &= 0: &\quad j &\geq B + 1 \\ \mu_j &= \mu; &\quad j &= 1, 2, \ldots, B + 1 \end{aligned} \qquad (44)$$

This system is stable with a throughput $\rho = \frac{\lambda(i,n,H)}{\mu} < 1$. Let us consider two separated cases: $\rho = 1$, and $\rho \neq 1$. For $j = 0,1,2,\ldots, B$ the distribution of the number of flows in the system is $P_j = \rho^j P_0$, which is determined according to

$$\begin{aligned} P_J &= \frac{\rho^j(1-\rho)}{1-\rho^{B+1}}; &\quad \rho &\neq 1 \\ P_j &= \frac{1}{B+1}; &\quad \rho &= 1 \end{aligned} \qquad (45)$$

Therefore, the rate at which the flows are blocked and lost is λP_B. The self-similar queuing process is described with the steady-state simulation procedure [16], presented in Figure 4.

The self-similar traffic can be generated and the sequence of arrivals is obtained. The fixed length of self-similar traffic is extracted by fixing the number of observations. As the service process is Markovian, the sequence of departures has exponential distribution, generated with an inverse transform generator [14].

The next step is the calculation of the buffer size. If the service size is greater than the size of arrivals, then the buffer size $B = 0$, as it is impossible to have a negative buffer size. In cases when the buffer size is greater than the overflow L, i.e. $B > L$, the traffic is lost, therefore we have made an assumption that $B = L$.

The simulation is performed with splitting of the sample path [15], using a variant based on the RESTART method [7], where any chain is split by a fixed factor when it hits a level upward, and one of the copies is tagged as the original for that simulation level. When any of those copies hits that same level downward, if the copy is the original it just continues its path, otherwise it is killed immediately. This rule applies recursively, and the method is implemented in a depth-first fashion, as

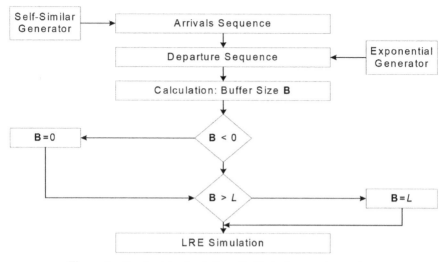

Figure 4   Steady-state simulation of self-similar queuing system.

follows: whenever there is a split, all the non-original copies are simulated completely, one after the other; then the simulation continues for the original chain [20].

The reason for eliminating most of the paths that go downward is to reduce the work. Therefore, the buffer size calculations being made for all sequences provide the opportunity to estimate the overflow probability using the steady-state simulation based on the RESTART method with the limited relative error (LRE) algorithm.

3.3 RESTART Method with Limited Relative Error Algorithm and Simulation Results

The limited relative error algorithm helps to determine the complementary cumulative function of arrivals at single server buffer queues with Markov processes. In order to describe the LRE princilpes for steady-state simulation in Discrete-Time Markov Chains (DTMC), let us consider a homogeneous two-node Markov chain, which is extended to regular DTMC, consisting of $(k + 1)$ nodes with states, respectively S_0, S_1,..,., S_k, as shown in Figure 5.

We obtain the random generated sequence x_1, x_2,..., x_t, x_{t+1}... for $x = 0, 1,..., k$, for which a transition for state S_j at the time t exists, e.g. $x_t = j$ and there are no constraints to the parameters of the transition probabilities:

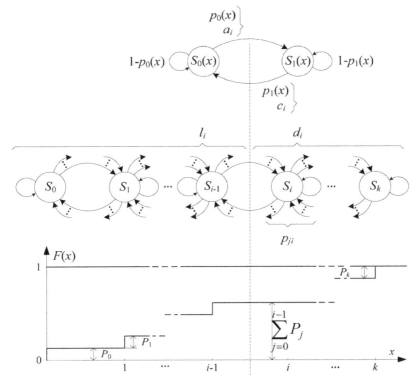

Figure 5 Cumulative function F(x) for (k + 1)-node Markov chain.

$$P_{ij} = P(j|i); \qquad (i,j) = 0,1,\ldots,k; \qquad \sum_{j=1}^{k} p_{ij} = 1 \qquad (46)$$

There are no absorbing states S_i, at $p_{ii} = 1$ for all stationary probabilities $P_{jj} = 0,1,\ldots,k$, which satisfy the constraint condition:

$$0 \le P_j < 1; \qquad \sum_{j=0}^{k} P_j = 1 \qquad (47)$$

The cumulative distribution *F(x)* can be presented as:

$$\left. \begin{array}{lll} F(x) = F_i; & (i-1) \le x < i; & i = 1,2,\ldots,k+1; \\ F_i = \sum_{j=0}^{i-1} P_j; & F_0 = 0; & F_{k+1} = 1; \end{array} \right\} \qquad (48)$$

In order to simulate the *(k+1)* nodes of Markov chain, the complementary cumulative distribution *G(x)=1–F(x)* that is more significant, can be determined along with the local correlation coefficient *ρ(x)* through the limited relative error approach.

After having the homogeneous two-node Markov chain defined as shown in Figure 3, with changing the states *n* times, an estimation of the local correlation coefficient $\hat{\rho}(x)$ can be obtained, which connects the number of transitions through a dividing line $a_i \approx c_i$, with the total number of observed events $l_i = n{-}d$, *(β = 0,1, ...i − 1,)* at left side, and d_i, at right side(*β = i, i + 2, ..., k*).

The value of simulated complementary cumulative distribution $\hat{G}_i$ can be defined directly by using relative frequency d_i/n, if there is enough number of samples:

$$n \geq 10^3; \qquad (l_i, d_i \geq 10^2); \ (a_i, c_i, l_i - a_i, d_i - c_i) \geq 10 \qquad (49)$$

The posterior equations can be used for the complementary function $\hat{G}(x)$, the average number of generated values of $\hat{\beta}$, the local correlation coefficient $\hat{\rho}(x)$, the correlation coefficient Cor[*x*] and the relative error RE[*x*]:

$$
\begin{aligned}
&\hat{G}(x) = \hat{G}_i = d_i/n \qquad \hat{\beta} = \frac{1}{n} \sum_{i=1}^{k} d_i \\
&\hat{\rho}(x) = \hat{\rho}_i = 1 - \frac{c_i/d_i}{1-d_i/n} \qquad i - 1 \leq x < i \\
&\hspace{5cm} i = 1, \dots, k
\end{aligned}
\qquad (50)
$$

$$\mathrm{Cor}[x] = \mathrm{Cor}_i = (1 + \hat{\rho}_i)(1 - \hat{\rho}_i) \qquad \mathrm{RE}[x]^2 = \mathrm{RE}_i = \frac{1-d_i/n}{d} \cdot \mathrm{Cor}_i$$

The main advantage of this approach is that the relationships between transitions c_i are obtained with routine statistical calculations. The necessary total number of simulation trails *n* is determined with the maximal relative error $\mathrm{RE}_{\max}[x]^2$ and with the less value of the function *G(x)*, presented as $G_{\min} = \hat{G}_k$ in approximation equation:

$$
\begin{aligned}
&n = \frac{(1-G_{\min})}{G_{\min}.\mathrm{RE}_{\max}[x]^2} \approx \frac{\mathrm{Cor}_k}{\hat{G}_k.\mathrm{RE}_{\max}[x]^2}; \\
&\mathrm{Cor}_k = \frac{1+\hat{\rho}_k}{1-\hat{\rho}_k}
\end{aligned}
\qquad (51)
$$

This method can be described with a standard version of limited relative error algorithm for random discrete sequences of buffer arrivals. It consists of the following three steps provided below.

Algorithm 3 Restart with Limited Relative Error Simulation Method

Step 1: Initialization of minimal and maximal values of the simulation parameter.

Step 2: Estimation and management of the simulation time.

Cycle L_1: Determine the current variable for calculating the Markov chain, e.g. $\omega := \beta$; generate a new value for β with given distribution.

Increase the number of state h_β.

If the condition $\beta < \omega$ is true, then increase the number of transitions $c_{\beta+1}$ while it reaches the value of c_ω.

Cycle L_2: Determine the total number of events at the left part l_s and at the right part d_s of the Markov chain and number of transitions $a_s := c_s$; check on the constraint condition (49) for the index $i = s$.

If the constraint condition (49) is true, then calculate the posterior values of the local correlation coefficient $\hat{\rho}_s$ and relative error $RE[x]$ with use of (50). Calculate whether the relative error $RE[x] \leq RE_{max}[x]$.

If $s < k$, than leave the cycle L_2.

If the index $s = k$ is reached, than leave the cycle L_1 and increase the index of the simulation time $s: = s+1$;

Step 3: Printing out the experimental results for $i = l, 2, ..., k$ The results for the total frequency d_i are determined according to (52):

$$d_i = \sum_{j=1}^{k} h_j \ \text{for} \ i = 0, 1, \ldots, k \ \text{where} \ d_o = n \tag{52}$$

The values of the complementary function $\hat{G}_i$, the local correlation coefficient $\hat{\rho}_s$ and the relative error $RE[x]$ are calculated as given in (50).

As an example, the overflow probability of an SSM/M/1/B self-similar queuing system has been simulated with different characteristics of long-range dependent self-similar arrival processes. In order to demonstrate the effects of self-similarity on the buffer overflow probability, the obtained experimental results were compared with the complementary cumulative distribution in the traditional single server finite buffer queue M/M/1/B. The obtained results in a logarithmic scale are given in Figure 6.

In order to get representative and steady results the sequences of 10000 observations were used. With the suggested LRE algorithm the values of complementary cumulative function $G(x)$ for different buffer sizes were calculated. The calculations were provided with the step $m = 4$. One can see in Figure 6 that the increasing Hurst parameter has led to an insignificant decrease in the overflow probability.

For example, for the value of Hurst parameter $H = 0.6$ the overflow probability was $G(L) = 1.045 * 10^{-1}$, and for $H = 0.9$ it was $G(L) = 5.6 * 10^{-2}$.

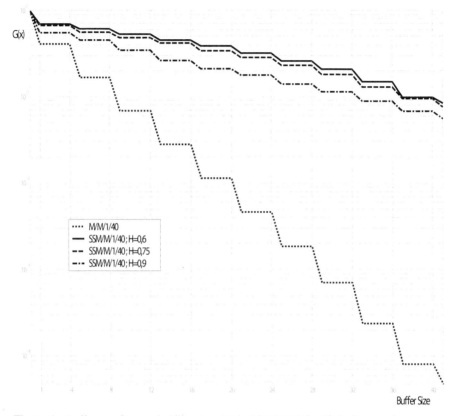

Figure 6   Buffer overflow probability (L = 41) in SSM/M/1/40 self-similar queuing system.

On the other hand, the overflow probability of self-similar queuing system has increased significantly in comparison with the theoretical M/M/1/B self-similar queuing system, for which $G(L) = 4.79*10^{-5}$.

After that, the simulation was repeated for SSM/M/1/B self-similar queuing system by using long-range dependent self-similar arrival process with $H = 0.6$ and different buffer sizes. The obtained results for buffer size $B = 40$, $B = 60$ and $B = 80$ are shown in Figure 7. One can see that since the buffer size was increased twice, the overflow probability has been changed simply by about two orders of magnitude – from $1.045*10^{-1}$ to $6.4*10^{-3}$.

Finally, it was confirmed that in order to design a single server finite buffer model with long-range dependent self-similar arrival processes, the buffer size has to be increased many times in order to decrease the overflow probability.

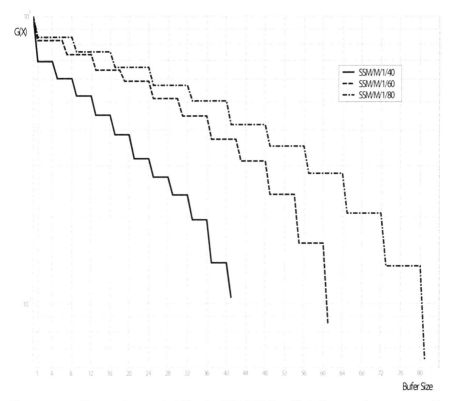

Figure 7 Buffer overflow probability in SSM/M/1/B self-similar queuing system with different buffer sizes.

4 Conclusions

The paper recommends new methods to estimate effectively the probability of buffer overflow in high-speed communication networks. The frequency of buffer overflow in queuing systems is very small; therefore the overflow can be defined as rare event and estimated using rare event simulation with Markov chains.

Initially, two-node queuing systems have been considered in this paper; and an event of buffer overflow at the second node was studied. Two efficient rare event simulation algorithms, based on the Importance sampling and Cross-entropy methods, have been developed and applied to accelerate the buffer overflow simulation with Markov chain modeling. Simulation results were shown and analyzed.

Then, steady-state simulations of self-similar queuing systems have been conducted using the RESTART method with Limited Relative Error algorithm to estimate effectively the probability of buffer overflow. The models of SSM/M/1/40 self-similar queuing system have been applied with different parameters of arrival processes and different buffer sizes. Simulations results were shown and analyzed.

The resulting recommended methods to estimate effectively the probability of buffer overflow are appropriate and particularly efficient being used for performance evaluation in high-speed communication networks, while higher performance networks must be described by lesser buffer overflow probabilities.

Acknowledgements

The author would like to thank her colleagues Michael R. Bartolacci from Penn State University – Berks, USA and Cees J.M. Lanting from CSEM, Switzerland for their time, thoughtful insights and review during the preparation of this paper.

References

[1] Bobbio, A., Horváth, A., Scarpa, M., Telek, M. "Acyclic discrete phase type distributions: Properties and a parameter estimation algorithm." *Performance Evaluation*, 54(1), 1–32, 2003.

[2] Bolch, G., Greiner, S., Meer, H, Trivedi, K. Queueing Networks and Markov Chains: Modeling and Performance Evaluation with Computer Science Applications, NY: John Wiley & Sons, 1998.

[3] Bueno, D. R., Srinivasan, R., Nicola, V., van Etten, W., Tattje, H. "Adaptive Importance Sampling for Performance Evaluation and Parameter Optimization of Communication Systems." *IEEE Transactions on Communications,* 48(4), 557–565, 2000.

[4] Bucklew, J. "An Introduction to Rare Event Simulation." *Springer Series in Statistics,* XI, Berlin: Springer-Verlag, 2004.

[5] C´erou, F., LeGland F., Del Moral P., and Lezaud P. "Limit theorems for the multilevel splitting algorithm in the simulation of rare events". *Proceedings of the 2005 Winter Simulation Conference*, San Diego, USA, 682–691, 2005.

[6] De Boer, P., Kroese, D., Rubinstein, R. "Estimating buffer overflows in three stages using cross-entropy." *Proceedings of the 2002 Winter Simulation Conference*, San Diego, USA, 301–309, 2002.

[7] Georg, C., Schreiber, F. "The RESTART/LRE method for rare event simulation", *Proceedings of the Winter Simulation Conference*, Coronado, CA, USA, 390–397, 1996.

[8] Giambene, G. *Queueing Theory and Telecommunications: Networks and Applications*. NY: Springer, 2005.

[9] Heidelberger, P. "Fast Simulation for Rare Event in Queueing and Reliability Models." *ACM Transactions of Modeling and Computer Simulation,* 5 (1), 43–85, 1995.

[10] Kalashnikov, V. Geometric Sums: Bounds for Rare Events with Applications: Risk Analysis, Reliability, Queueing. Berlin: Kluwer Academic Publishers, 1997.

[11] Keith, J., Kroese, D.P. "SABRES: Sequence Alignment by Rare Event Simulation." *Proceedings of the 2002 Winter Simulation Conference*, San Diego, USA, 320–327, 2002.

[12] Kroese, D., Nicola, V.F. "Efficient Simulation of a Tandem Jackson Network." *Proceedings of the Second International Workshop on Rare Event Simulation RESIM'99*, 197–211, 1999.

[13] Lokshina, I. "Study on Estimating Probability of Buffer Overflow in HighSpeed Communication Networks." *Proceedings of the 2014 Networking and Electronic Commerce Conference (NAEC 2014*, Trieste, Italy, 306–321, 2014.

[14] Lokshina, I. "Study about Effects of Self-similar IP Network Traffic on Queuing and Network Performance." *Int. J. Mobile Network Design and Innovation,* 4(2), 76–90, 2012.

[15] Lokshina, I., Bartolacci M. "Accelerated Rare Event Simulation with Markov Chain Modelling in Wireless Communication Networks." *Int. J. Mobile Network Design and Innovation,* 4(4), 185–191, 2012.

[16] Radev, D., Lokshina, I. "Performance Analysis of Mobile Communication Networks with Clustering and Neural Modelling". *Int. J. Mobile Network Design and Innovation,* 1(3/4), 188–196, 2006-a.

[17] Radev, D., Lokshina, I. "Rare Event Simulation with Tandem Jackson Networks." *Proceedings of the Fourteen International Conference on Telecommunication Systems: Modeling and Analysis – ICTSM 2006,* Penn State Berks, Reading, PA, USA, 262–270, 2006-b.

[18] Radev, D., Lokshina, I. "Advanced Models and Algorithms for Self-Similar Network Traffic Simulation and Performance Analysis", *Journal of Electrical Engineering,* Vol. 61(6), 341–349, 2010.

[19] Rubino G, Tuffin B. *Rare Event Simulation using Monte Carlo Methods*, UK: John Wiley & Sons, 2009.

[20] Villen-Altamirano, M., Villen-Altamirano, J. "On the efficiency of RESTART for multidimensional systems." *ACM Transactions on Modeling and Computer Simulation*, 16 (3), 251–279, 2006.

Biography

I. Lokshina, PhD is Professor of Management Information Systems and chair of Management, Marketing and Information Systems Department at SUNY Oneonta. Her positions included Senior Scientific Researcher at the Moscow Central Research Institute of Complex Automation and Associate Professor of Automated Control Systems at Moscow State Mining University. Her main research interests are complex system modeling (communications networks and queuing systems) and artificial intelligence (fuzzy systems and neural networks).

Author Index

Keywords Index

www.ingramcontent.com/pod-product-compliance
Lightning Source LLC
LaVergne TN
LVHW012333060326
832902LV00011B/1870